Fuzzy Logic

About the Author

Lefteri H. Tsoukalas is a professor at Purdue University, where he has taught fuzzy logic for nearly three decades. Dr. Tsoukalas is the recipient of numerous honors and awards, including several best teacher awards and the *Humboldt Prize*, Germany's highest honor for international scientists. He is *Fellow* of the American Nuclear Society.

Fuzzy Logic

Applications in Artificial Intelligence, Big Data, and Machine Learning

Lefteri H. Tsoukalas

New York Chicago San Francisco
Athens London Madrid
Mexico City Milan New Delhi
Singapore Sydney Toronto

Fuzzy Logic: Applications in Artificial Intelligence, Big Data, and Machine Learning

1 2 3 4 5 6 7 8 9 LHN 29 28 27 26 25 24 23

Library of Congress Control Number: 2023943898

ISBN 978-1-264-67591-3
MHID 1-264-67591-7

Sponsoring Editor
Lara Zoble

Editorial Supervisor
Janet Walden

Project Manager
Tasneem Kauser,
KnowledgeWorks Global Ltd.

Acquisition Coordinator
Olivia Higgins

Copy Editor
Girish Sharma

Proofreader
Rup Narayan

Indexer
Lefteri H. Tsoukalas

Production Supervisor
Lynn M. Messina

Composition
KnowledgeWorks Global Ltd.

Illustration
KnowledgeWorks Global Ltd.

Art Director, Cover
Jeff Weeks

To
Vivi **and** *Kostas*

Contents

Preface

This book is based on three decades of teaching a course in Fuzzy Logic at Purdue University. The course has drawn undergraduate and graduate students from a variety of disciplines including, but not limited to, Engineering, Computer Sciences, Mathematics, Agriculture, Economics, and Social Sciences—some attacted by novelty, others by intellectual curiosity, most by the need to create models for research in systems suspected of nonlinearities and complexities. Almost all students are initially surprised by the idea that language, the language we use in everyday life, may be viewed as a form of computation. They are also surprised to learn that fuzzy logic is not an oxymoron, and that fuzzines enables artificial intelligence (AI) to become *actionable*, *interpretable*, and *tranferable*. The surprise gives way to enthusiasm when class projects turn theoretical ideas about fuzzy *if/then* rules, relations, inference, and compositions into practical success with decision-making and control algorithms.

Although student enthusiasm with fuzzy logic may be the outcome of practical success, the importance of the field rests on growing interest in AI as computers approach and possibly exceed the limits of Moore's Law. For instance, 2 nm chips hosting nearly 50 billion transistors in an area the size of a fingernail promise widespread use of machine learning (ML) to process Big Data and enhance the capabilities of individuals and organizations to provide and use reliable intelligence about the future.[1]

In addition, large language models used by chatbots simulating human-like machine dialogues promise to go beyond novelty to accelerate innovation in productivity-driven areas, including operations and maintenance, advanced manufacturing, and logistics. Fewer people will be able to accomplish more in the productive machinery of society through AI-enabled collaboration and coordination across diverse

[1]For comparison, the 1950 Manchester Computer of Alan Turing had a total memory capacity of 1280 bits, while modern laptops may have 64 GB RAM and 4 TB or more storage memory.

domains and scales of time, space, and complexity. In addition, emerging AI capacities for growing knowledge through large-scale purposeful activities may produce disruptive transformations in research, education, security, business, health-care, entertainment, and possibly even governance.

Of course, there are the enduring philosophical questions. What do they have to do with AI? Evidently, they concern not only AI from a theoretical perspective, but also engineering AI, that is, making AI useful—to delineate its boundaries, to advance its utility, to bring it to where it is *actionable*, *interpretable*, and *transferable*.

In this book, we place fuzziness at the epicenter of a linguistic view of AI, a view that draws upon philosophical questions, for philosophy's original meaning of "love of inquiry" is also the meaning of scientific research. For AI to be actionable, interpretable, and transferable, we revisit the old philosophical discourse on reason and language. It is a discourse where meaning is observed and judged against a social linguistic background along the lines of the famous *imitation game* of Alan Turing. In his seminal 1950 publication "Computing Machinery and Intelligence," Turing gave us the basis for judging intelligence and therefore learning in an actionable, interpretable, and transferable way. In Chap. 1, we delineate the framework of fuzziness, language, and AI, while in Chap. 6 we examine a modern version of Turing's imitation game using a digital twin for Purdue's reactor, the PUR-1, the first fully digital instrumentation and controls (I&C) nuclear reactor in the United States.

The old *Mind-Body duality* of Plato and Descartes may provide boundaries for AI, often viewed as an intrinsic property of machines, rather than a feature of human-machine interactions. If, as Plato thought, the mind is the seat of intelligence and the body is a mere vessel for the mind, it may follow by analogy that the computer is a vessel for AI. This is a view that is intuitively correct but actually of limited value.

In this book, we embrace an alternative view which holds that AI, like language, is essentially a social phenomenon. Wittgenstein in his *Philosophical Investigations* (1952) argues that a private language is an impossibility, since we cannot have a meaningful language that is only understood by one-and-only-one person. Therefore, language is an inherently public and shared activity. In Chap. 1, we examine these ideas through a thought experiment, where 7000 speakers, each using one and only one language, find themselves on a desert island. The *Desert Island Thought Experiment* is a stage for introducing the Wittgensteinian idea of "language-game" which provides a launching pad for fuzzy algorithms. In Chaps. 2 and 3, we lay out the mathematical elements for fuzzy logic computations, and in Chaps. 4 and 5 we put these ideas to use with language-games for decision-making, control, and forecasting. Supporting materials, including review questions and homework problems, as well as color versions of the appendices with the Python code used in the examples in Chaps. 4 and 5, are provided on the McGraw Hill Professional Web site book landing page at mhprofessional.com/tsoukalas.

As Wittgenstein recognized, language requires conventions formulated and shared by groups. These conventions can become fuzzy *if/then* rules, thus transforming language into computational processes. A private language, on the other hand, would be a language that is known only to one person and is not shared with anyone else. Such a language would not have any public rules or conventions, and therefore cannot be actionable, interpretable, or transferable.

If it is not possible for an individual to create a private language with its own meaning, independent of any social or cultural context, one may argue that the mind-body duality is a false dichotomy. This thesis was put forward by the eminent British philosopher Gilbert Riley (1900–1970) in his seminal work *The Concept of Mind* (1949), where he critiqued the mind-body duality as a *category mistake*, as for instance, that which is made by a visitor to Oxford who after viewing the library and colleges reportedly inquired, "But where is the University?" The visitor's mistake is presuming that a university is part of the category "units of physical infrastructure," rather than that of an "institution."

The view that AI is *not* an intrinsic property of machines but a phenomenon evolving against a social linguistic backgroung, transends the mind-body duality and focuses instead on *shared intentionality*, that is, the uniquely human quest to read the minds of others, a quest that leads to a concept of *understanding* which is social and not private. For instance, when a student says "I understand," he is not reporting a private experience of understanding because the criterion for understanding is not his own claim, but rather a teacher's verification. A teacher decides whether or not he really does understand, say, differentiation and integration. If the student can apply the rules of calculus to differentiate and integrate, a teacher considers that the student has reached a commensurate level of understanding. To understand something is simply to be able to apply it. It is not private experience but public or external judgment that affirms understanding.

Thus, AI may facilitate the development of understanding by becoming part of a public linguistic background that affirms and guides learning. In this book, we approach fuzziness as an indispensable methodological enabler that makes AI an *actionable*, *interpretable*, and *transferable* element in an evolving social linguistic background.

West Lafayette, March 2023

Acknowledgments

Over a rather long period of time many individuals have contributed to the making of this book, hence any list of acknowledgments is bound to be pitifully incomplete. Nonetheless, special thanks are due to professor Seungjin Kim, head of the School of Nuclear Engineering at Purdue and my colleagues, professors H. Abdel-Khalik, M. Bertodano, H. Bindra, S. Chatzidakis, C. K. Choi, A. Garner, A. Hassanein, M. Ishii, X. Lou, S. Revankar, R. Taleyarkhan, Y. Xie, and Y. Xu. Deep gratitude is due to colleagues and university leaders who facilitate the *Center for Intelligent Energy Systems* (*CiENS*) that enabled the book, including Purdue President Mung Chiang and Deans Arvind Raman, Mark S. Lundstrom, and Weinong Chen. Thanks are due to past and current members of the *AI Systems Lab* (*AISL*) who contributed, reviewed, and tested material including Drs. A. Bougaev, R. Gao, Vivek Agarwal, as well as S. Pantopoulou, R. Appiah, K. Prantikos, M. Pantopoulou, and U. Cotul. Special thanks to professors D. Bargiotas, A. Chroneos, A. Daskalopulu, A. Heifetz, G. Stamoulis, and M. Vassilakopoulos for critical input. I would like to thank the Krach Institute for Tech Diplomacy at Purdue, and in particular Zachary Goldsmith, whose award of a Senior Research Fellowship has allowed me to continue work on fuzzy logic in a new frontier, the intersection of technology and foreign policy. Many scholarly activities of *AISL* that found their way into the book were enabled by the intellectual and material generosity of two extraordinary Purdue alums, Dr. Kostas Pantazopoulos (Class of 1998) and Mrs. Vivi Galani (Class of 1996). They have been a constant source of inspiration and I am deeply grateful for their support. Many thanks are due to Tasneem Kauser and the outstanding professionals of KnowledgeWorks Global, Ltd. for their exemplary art in reaching the final book form. Lastly, but most importantly, this project would not have been possible without the good people at McGraw Hill, especially Lara Zoble, whose professionalism and patience turned a fuzzy proposal into a book.

CHAPTER 1

Language and Computation

Where the role and function of linguistic descriptions in human-machine communications, human collaborations, and societal organization is examined.

The emergence of digital computers forced mathematically inclined engineers to come to terms with the idea of *language as a form of computation*. Prior to this, investigations regarding truth, logic, grammars, and semantics were found in the works of philosophers and linguists including preeminent scholars like Gottlob Frege, Bertrand Russell, Ludwig Wittgenstein, Jan Łukasiewicz, and Noam Chomsky. Their ideas inspired Lotfi A. Zadeh, in the mid-1960s, to generalize the notion of set membership and extend Łukasiewicz's *n-valued logic* to *infinite-valued logic* (Zadeh, 1965). At the time, very little attention was given to this development, but gradually it stimulated interest on the question of whether and how we can compute with words. Two decades later, Zadeh outlined a fuzzy logic approach to making computers think like people (Zadeh, 1984). After all, we use language to solve math problems, build models, write computer codes, investigate findings, interpret, revise, go through the labors of peer review, and place new knowledge in the bibliography. Why not use language directly for computing, why not make computers think like humans?

One impediment is that when we consider words in everyday life, they are useful as much as they are elusive. Established rules for the order of words, what we call *syntax*, often disguise or obfuscate the meaning of words, what we call *semantics*. To overcome the confusions arising from the conflict between syntax and semantics, we need to examine language at work, especially in activities of everyday life increasingly enabled by the use of AI (Artificial Intelligence). In a broad sense, we view AI as located within the milieu of natural language, as part of language at work. Conversely, considering language as a form of computation offers AI limitless possibilities for growth with great returns. This is the perspective taken in our book: Language and AI form a dyadic connection, with fuzzy logic serving as an important enabler of this relationship.

Zadeh, an erudite polyglot, held that if machines are to communicate better with humans, it is imperative to account for a phenomenon he called *fuzziness*. At first, fuzziness looks like a property of definition, what the linguists call *epistemic uncertainty*. Upon closer examination, however, fuzziness appears to be related to *parsimony*—the economy of words through which we can think, and do more with less. Fuzzy logic offers means for managing parsimony, the control of which is indispensable for the relation of language to reality. Zadeh's famous dictum "as complexity rises precise statements lose meaning and meaningful statements lose precision" can be thought of as a modern, computer-age articulation of the ancient principle of parsimony which is known in philosophy as *Occam's Razor*.

Fuzziness is found in the construction of groups or sets when we place within their linguistic perimeter elements with varying degrees of memberships, and thus, it is hoped, with a plethora of meanings. For instance, a small number of words for color such as *green*, *blue*, and *red* suffice to describe the millions of colors we can see and paint with. There is nothing fuzzy about the colors themselves or for this matter about anything in the *World*, out there. Things are what they are: planes, planets, suns, desks, or chairs. Fuzziness is an indispensable feature of the *symbolization of the World*, how a mind (or a computer) uses symbols like words, numbers, sets, categories, relations, implication, and inferences to engage, successfully, with the *World*.

"What is the meaning of a word?" is typically a question for which we need to consider a number of questions outside the question of meaning, a process which may give us important boundary conditions in AI. For instance, how is the meaning of a word learned, when do we know if someone understands the meaning of a word, what conventions or standards are needed, why and how are these conventions established, etc.? If we approach the question in this way, we see that to talk about the meaning of a word is to talk about the way in which the word is used. To say that a human, or a computer, understands the meaning of a word is simply to say that he/she/they/it has learned or understands how the word is used. The identification of *meaning* with *use* is important, because the meaning of a word is much more than an object for which the word may stand, and although some words do stand for objects, these words are a rather small fraction of language as a whole. We all use language daily and often without paying any particular attention to it, a practice called *ostensive definition*, in which we correlate words with objects by pointing to them and christening objects with words uttered at the moment of pointing. The act of pointing has important computational value. It is what computer peripherals are to computing, for instance a keyboard in relation to a desktop. It can be used to ask questions to name objects or actions, such as eating or walking, to introduce new words, but it can also be used to give orders. Hence the complexity of linguistic operations. As Wittgenstein would have pointed out, before pointing can give meaning to a word it must itself be understood as having meaning.

To anchor AI within a language framework, we emphasize the importance of action facilitated by language. In other words, we assert that *to know is to be able to do* or as the Italian renaissance philosopher Giambattista Vico (1668–1744) famously said *Verum est Factum*, meaning the truth is found in the action. Hence, the problem of computing with words may be approached as an inverse problem, a problem for which we know the outcomes and need to discover the causes that produce these results. An emphasis on action as an outcome from which we go backward to discern causes can make AI actionable, correctible, and interpretable, in the sense that actions such as correction and explanation are important for AI programs to learn and grow without forgetting what was previously learned and with some capacity to explain its history and future on demand to a human user, literally a human interlocutor with an AI program.[1] Zadeh appears to have envisioned fuzzy logic from its inception as a tool for enabling actionable, correctible, and interpretable AI (Zadeh, 1984).

To clarify these ideas, let us consider a *thought experiment*. Suppose we place on a desert island a speaker from each one of the 7000 languages spoken in the world today. Let's assume that each speaker speaks and understands only one language. Thus, we have on the island 7000 speakers, none of whom can use anyone else's language. What will happen in this island of 7000 different languages? What will it be like a day later, a week later, a month or a year later?

If no speaker can speak any language except his own, it is easy to imagine that at first, there will be chaos—7000 speakers of 7000 different languages in a situation like the biblical *Tower of Babel* in reverse. Since they cannot go their own ways, as in the Bible, they have to stay and survive on the desert island. How can they guess each other's intentions? Can they learn to communicate using ostensive definitions, gestures, pantomime, and facial expressions?

In the desert island thought experiment, there is absolutely no certainty about what may happen, but given enough time, possibly more than a week or a month, some common words for survival may get established and a mode of communication may emerge in

[1]In 1950, Alan Turing laid the foundations of AI in a seminal publication in the journal *Mind* titled "Computing Machinery and Intelligence" in which he paraphrased the question "can machines think?" to what he called an "imitation game," which is widely known as the Turing test, the essence of which is whether a machine can imitate human intelligence to a degree that may pass for a human. It is interesting to note that Turing considered the question "can machines think?" too meaningless to deserve discussion because we do not know what thinking is.

individuals and groups collaborating with each other. To survive they will discover action through words and words from action, words acquiring meaning through pointing at things and acting with things, words that label categories and describe conditions and consequences. But to do all these, something more is needed and this is *shared intentionality*, which roughly means: *reading each other's minds.*

In a series of seminal works, the preeminent anthropologist Michael Tomasello and his colleagues argue that *shared intentionality* is a uniquely human trait. Tomasello presents compelling evidence demonstrating an innate human desire to read the minds of others (Tomasello, 2014). According to Tomasello, this is the principal mechanism that facilitates the emergence of language. By looking at others (literally, tracking their gaze) and following their fingers pointing, or their bodies moving, words and rules about words get established and acquire a shared meaning. Words are cultural artifacts produced and reproduced in a never-ending social innovation process through which objects, activities, and states of mind, like happy or sad, can be shared by individual minds trying to coordinate a potentially infinite number of activities. Words are tools that outlive individuals with finite lives. As we discuss later in this chapter, no words are strictly private; they need the milieu of language, which is essentially a social linguistic background in order to become meaningful.

Tomasello's work provides compelling empirical evidence validating Wittgenstein's views on language found in the *Philosophical Investigations* (Wittgenstein, 1953) where every human activity, including those necessary for survival, involves some kind of "*language-game.*" This is a term that Wittgenstein coined to refer to language innovations for collaboration and coordination to accomplish a task. The correctness of language-games may not be arbitrated by grammatical or syntactic criteria (as Wittgenstein had claimed in the 1920s in his seminal book the *Tractatus*), but whether or not some desirable goal is accomplished (Wittgenstein, 1922). Thus, *meaning is use*, and by looking at use, we go back to infer meaning. For example, a bricklayer and an apprentice may invent a language-game for the purpose of bricklaying. The correctness of this language-game should be judged by whether they can build a wall. It is conceivable that bricklayer and apprentice may not even share the same natural language. But if they can build a wall, it proves that they have a language for it. The desert island speakers, likewise, may invent language-games for survival and hence establish words and meanings, which may flow from one language-game to another. For example, a language-game for discovering potable water may provide words whose meaning may flow over to another language-game aiming at discovering eatable fruits or facilitating the task of igniting a fire, of finding fuel (burnable woods), of establishing shelter, and even entertainment, education, and rest. Language in everyday life entails use of language-games with many similarities and a multiplicity of uses, which overlap and share words that may be used in many different language-games as well. In this sense Wittgenstein's view of natural language is that of the *power set* (the set of all sets) of all language-games, including formal languages such as axiomatic systems, logic, and computer languages.

To say that a language emerges as the solution to the inverse problem, for instance, to promote survival, leads us to language-games, which place emphasis on results produced by action, not the rules of syntax alone. Again, we evaluate language-games by looking at the accomplished goals and the shared intentions to arrive at these goals. To develop actionable, transferable, and interpretable AI, we look at language-games involved in the success of collaborative activities.

Language-games of collaboration point toward the relation between thought and reality and specifically toward the group of games needed for *constructing* the *symbolization of reality*.[2] In this sense, endowing sets with multiple memberships can be thought of as enabling a language-game of coordination. In fact, it is a game upon which most other language-games rest. If a constructive relation between language and reality is achieved, that is, between computing and doing, the language-games involved are deemed correct. This is a pragmatic view that may be objected to by some, but it has important engineering ramifications as it may provide metrics for success. It is a view that was treated with skepticism when fuzzy logic first appeared. But, it is worth recalling that similar things have happened in the past, often when a major paradigm shift is involved.

For instance, in 1734, the preeminent mathematician George Berkeley rejected calculus as mathematically incorrect and published an entire book titled *The Analyst; or, A Discourse Addressed to an Infidel Mathematician* as polemic treatise against calculus (Fig. 1.1). The infidel mathematician was none other than Sir Isaac Newton. Berkeley and many others attacked the basic notions of calculus as false because they were convinced that it was in conflict with Euclidean geometry. What they completely missed was that to model motion, we needed mathematics for the concept of *nearness* (which did not exist in Euclid).

As professor Jerome H. Manheim points out in *The Genesis of Point Set Topology*, "Berkeley's valid objections to the processes of

[2]The great German philosopher Ernst von Glasersfeld (1917–2010) and many others put forward a set of ideas called *radical constructivism* in order to bridge the gap in our understanding of how knowledge is created through education (Glasersfeld, 1987). In a similar spirit, the Austrian-American engineer Heinz von Foerster (1911–2002), a pioneer in the search for AI technologies, established in the 1950s the influential Biological Computing Laboratory at the University of Illinois at Champaign-Urbana (Foerster, 1984).

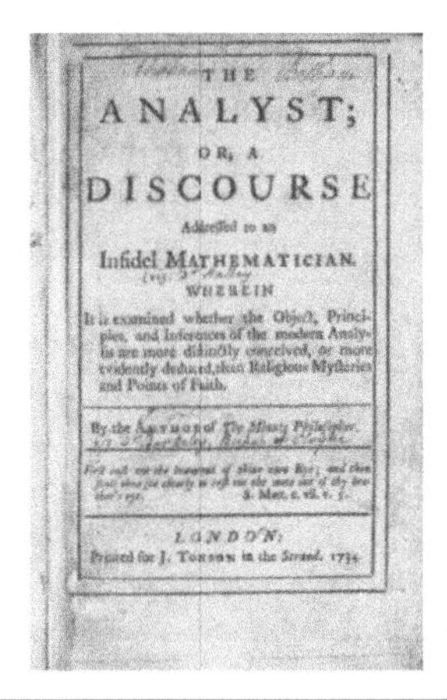

THE
ANALYST;
OR, A
DISCOURSE
Addressed to an
Infidel MATHEMATICIAN.
WHEREIN
It is examined whether the Object, Principles, and Inferences of the modern Analysis are more distinctly conceived, or more evidently deduced, than Religious Mysteries and Points of Faith.

By the AUTHOR of The Minute Philosopher.

First cast out the beam out of thine own Eye; and then shalt thou see clearly to cast out the mote out of thy brother's eye. S. Matt. c. vii. v. 5.

LONDON:
Printed for J. Tonson in the Strand. 1734.

FIGURE 1.1 Berkeley's 1734 treatise rejecting calculus as mathematically unsound titled, *The Analyst; or A Discourse Addressed to an Infidel Mathematician.*

Newton and Leibniz were not supplemented with a constructive program. The only possible constructive program, the *arithmetization* of analysis would have been most unlikely as long as (Euclidean) geometry occupied the dominant role" (Manheim, 1964). Focusing on the concept of "nearness" and its importance in physics enabled Newton and Leibnitz to establish calculus by extending, not refuting, Euclidean geometry. Analogously, Zadeh's focus on the concept of "*fuzziness*" is an absolute necessity for understanding language as computation and engineering new tools for AI to harvest the ever-growing sophistication of computers.

The mathematics of fuzziness are at once profound and simple. In fuzzy sets the characteristic function $\chi_A(x)$ of a set A is extended to a membership function $\mu_A(x)$, simply by changing the valuation set of A from a two-element set, that is the set $\{0, 1\}$, to a set with infinite elements, that is the interval from 0 to 1, or $[0, 1]$. Mathematically, this is indicated by changing the characteristic function

$$\chi_A(x): X \rightarrow \{0, 1\}, \quad x \in X \tag{1.1}$$

to a membership function

$$\mu_A(x): X \rightarrow \ [0, 1] \ , \quad x \in X \tag{1.2}$$

In Eq. (1.1), the characteristic function is the mathematical identification of the set A. It answers the question "Does x belong to A?" There are only two ways to answer: Yes (labeled 1) or No (labeled 0). In Eq. (1.2), there are many, possibly infinite answers. Zadeh's point of departure was Łukasiewicz's *n-valued logic,* but his destination was a *theory of commonsense reasoning* that aims to represent the whole of human knowledge in a logical system in which everything is derived from fuzzy sets as mathematical descriptions for words and rules of inference ranging from a minimum initial stock of basic propositions and inferences to compound terms and elaborate reasoning schemes. In addition to generalizing Łukasiewicz's *n-valued logic* to infinite valued logic, Zadeh used three features initially found in *Principia Mathematica* (Whitehead, 1925):

(i) The *principle of extensionality* which asserts that mathematical relations can be true even when their input variables are fuzzified. In other words, fuzzy generalizations do not lead to nonsensical statements and hence compound propositions can be composed from simpler ones. This leads naturally to *if/then* rules and their aggregations, called *fuzzy algorithms*, very useful in a variety of AI applications including, but not limited to, modeling, decision-making, diagnosis, man-machine communications and control.

(ii) Using *exceptions*. Fuzzy algorithms are linguistic descriptions, in which partially true or false expressions are strengthened or refuted through exceptions. Fuzziness provides a natural way of using exceptions as a method for strengthening assertions to achieve parsimonious descriptions.

(iii) Connecting *syntax to semantics*. Syntactic limitations that lead to semantic nonsense or logical paradoxes where grammatically well-formed sentences could nevertheless be meaningless, are elucidated by fuzziness. Distinctions between true and false seek nonsyntactical criteria (that is results related to action), which is something that allows fuzziness a new role in a *theory of meaning* that distinguishes between the significant and the senseless.

What, then, is meant by equating truth with action? If it is true that doing can help to improve our knowing, the phenomenal success of fuzzy logic is definitely found in the doing. The intelligent performance of some action undoubtedly promotes one's understanding of it. It is evident that this occurs most naturally in the field of social action, but the same can be held for AI in engineering applications. If a system is successful in performing some action and in the process it can also improve, then we say that the system *learns by doing*. For instance, if an AI program shadows an individual in sufficiently detailed but also diverse domains, the AI program can reach a level of competence about the individual's biography including preferences, relations, credentials, history, expressions, and interests so that the AI program can facilitate a kind of dialogue with the individual, even when the individual no longer exists. To place AI in the milieu of language opens the prospect of achieving a kind of "immortality." Our physically limited being may survive in memory (if we so desire) in sufficiently rich and accurate detail to facilitate virtual dialogues unconstrained by space and time limitations. Of course there are important issues of privacy and free-will and all needs to be done within institutions that respecting individual rights and preferences, but it is apparent that the possibility emerges of greater knowledge transfer amongst generations and possibly an easier and even more useful accumulation of beneficial wisdom for all.[3]

Fuzzy logic exploits as much as facilitates the computational nature of language as an instrument for action. Rules for action use natural language where affirmations, exceptions, correlations, and causation can all be tested, validated, and strategically integrated. Through numerous engineering successes, fuzzy *if/then* rules and algorithms make computers work in ways that we humans can say, in praise or scorn, that computers think like humans. Besides pragmatic considerations, there are some fundamental theoretical reasons that account for this success. In the 1990s, Nguyen, Kosheleva, and Kreinovich have shown that the formal concept of logical equivalence holds for fuzzy logic (Nguyen et al., 1996). This can be checked algorithmically and the checking is done by showing that a strong equivalence notion coincides with the equivalence in logic programming. Two very different approaches, fuzzy logic and logic programming, lead to the same notion of equivalence between proposition formulas. Hence, the success of fuzzy logic is not really paradoxical. It is due to the idea that language and hence thought is a form of computation and computers can be made, to a degree, to think like people.

Computationally, fuzzy logic encodes *facts* that are distinct from and independent of one another and can be articulated by elementary propositions such as

$$p^1 = x \text{ is } A$$

and

$$p^2 = y \text{ is } B$$

where, x and y are fuzzy variables, and A and B are fuzzy values for these variables, each possessing its own membership function.[4]

Elementary propositions such as those above owe their meaning and truth not only to their relation to other propositions but, most importantly, to the relation with the *World*. Fuzzy propositions are simple and have a definite relation to action and not merely stand in internal logical relations to one another; their meaning is derived from experience.

---

[3]Large language models (LLM), such as used by popular bots like *chatgpt*, use pretrained sets of connections between words to take inputs and deliver as outputs the prediction of sentences and phrases. Such predictions can be made statistical AI tools, referred to as generative pretrained transformers, within domains that define language-games the boundaries of which need to be determined by AI engineers making judgments encodable as *if/then* rules.

[4]The fundamental mathematics and operations of fuzzy logic are found in Chap. 2.

Rules of the *if/then* variety are associations of elementary or compounds propositions, the latter can be generated from the former by the operations of *negation* (the symbol of which is – or ¬ and corresponds to the logical operator *NOT*), *conjunction* (the symbol for conjunction is ∧ and logically it corresponds to the connective *AND*), and *disjunction* (the symbol for disjunction is ∨ and logically it corresponds to the connective *OR*). To assert a compound proposition is to assert a collection of elementary propositions. Thus, the logical operators *NOT, AND, OR*, and the rules "*if/then*" are for the convenient assertion and association of elementary propositions, the ultimate bearers of veracity.

How do sentences such as *if/then* rules fail, that is, how do they become false or even nonsensical? Fuzzy logic answers such questions relying not on theory, *but on activity*. The kind of analysis that provides verification is based on circumstances and it is not limited by thinking of language as a kind of logical calculus. Language in certain technical areas can be logically structured in the manner of a calculus, but the elasticity of language is what makes it possible to put old language-games to new uses. In the thought experiment of the desert island, the language-game of seeking potable water will likely provide words and phrases that may spill over to other language games, for instance language-games for finding eatable fruits, finding safe shelter, or even entertainment.

To access the minds of others, AI needs a lot more than a narrow logical calculus; it needs to report and describe mental states so as to make them computable. Tomasello's shared intentionality implies that mental states are shared through words derived from publicly observable situations, that is, within a social environment or what may be described as a social linguistic background. A *private language*, referring to the experience of which only one person is aware, *is in this sense an impossibility*. For instance, when someone says "I understand," he is reporting a private experience of understanding. But whatever experience he may have, what decides whether or not he really does understand, say arithmetic, is whether or not he can go on to use arithmetic. Understanding something is established against a social background; to wit, a correct application of arithmetic has to be publicly evidenced, for instance, by a teacher in a teacher–student context where a teacher evaluates the application or arithmetic and can show if the interpretation made of it is the correct one. Essentially the same argument is applied to the concept of meaning something by a word. What a person or an AI program means by a word is not a private experience. The meaning attached to a word is revealed by considering the things to which it is applied, and from which it may be withheld. Words and verbal contexts, statements, and arguments acquire meaning, verification, and validity only against a social background. It follows from this that thinking is not an interior or solipsistic process that accompanies speech.

The same general treatment is extended to cover emotions such as "hope" and "fear," which derive their significance from the social linguistic background, that is, the surroundings of the people to whom they are ascribed and not to some private events going on within them. An important feature of the social linguistic background is what the people to whom they are ascribed will say. The AI machines mediating people's thoughts and language are tools facilitating human dialogues, and as alluded earlier, where these dialogues may no longer be as limited by space and time constraints since AI enables human–machine exchanges. It should be noted, however, that they are tools functioning against a human (social or cultural) background. As Wittgenstein pondered, what sense is there to the phrase that a dog is afraid of something that may happen next week?

A social linguistic background provides the criteria for our correct use of words and AI may use such criteria to be actionable, transferable, and interpretable. Since there cannot be a language whose use is wholly determined by private experience, AI *is an essentially social phenomenon as is language*. In as much as the making of noises does not become linguistic utterance unless it is governed by rules, AI programs do not learn, grow, correct, or self-explain unless there is an applicable distinction between the correct and mistaken use of words. With a private language, this condition cannot be satisfied, and the uttering of words introduced as names of private sensations could simply be described as an "empty ceremony" (Pitcher, 1966). It is for this reason that our mental words must be, as they are, connected with features of our situation which anyone can in principle observe. Every inner process must have its external affirmation.

The theory of shared intentionality implies that intention, motivation, volition, and even free-will are no more private and internal than thoughts and feelings. There is no mental language with which we could talk about our own mind unless there is a public mental language and we had mastered it. A set of metrics AI can be based foremost on the realization that there is no exclusive private language, and hence develop tools for measuring the growth and progress of AI systems. A performance metric would enable us to initially compare, first, different AI systems against the same social linguistic background, and second, the same AI system against different linguistic backgrounds. AI located against social linguistic backgrounds leads us to the following five guidelines for actionable, transferable, and interpretable AI systems.

Guideline 1: General vs Specific. General metrics lead us to idealizations similar to crisp sets as opposed to fuzzy, and hence they may not be suitable for AI working in some specific domain where application-oriented metrics may be more useful.

Guideline 2: Generalization. This metric has to do with the ability to generalize, that is, to compute inductively. In many ways, AI efficacy is measured by the capacity for generalization. However, generalization can happen at different levels as illustrated in Fig. 1.2 where at the low level we see four instances of geometric objects: a triangle, a rectangle, a line segment, and a circle. If an AI program were to recognize them for what they are, it could use a direct matching approach, where the seen instances have to be exactly or directly matched with what is in memory. Alternatively, memory can have generalized instances of these object where some topological transformation such as a segment of a curve can be used to indirectly match a line segment or going even further in abstraction we can have composite generalized instances that include in the form of a rectangle that can be stretched to become a circular or oval object, or a line segment which is included in memory as the jingled side of a triangle.

Guideline 3: Adaptability. This guideline has to do with learning so that the AI system can transfer learning from one domain to another; initially the two domains can be close enough but different and later transfer to a third domain and so on and so forth (see the transitivity Guideline 5). Hence, our system uses language-games that are related but different and a cross-over of words and rules from one language-game to another. If the domain or relevant language-game changes drastically, AI should be able to report limitations in adaptability to avert the emergence of nonsense at the expense of adaptability.

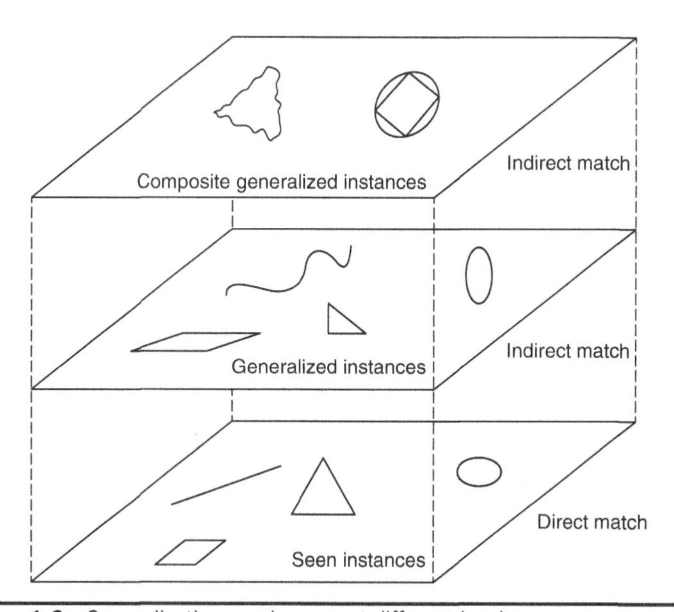

FIGURE 1.2 Generalization can happen at different levels.

Guideline 4: Metrics on the Social Linguistic Background. Digital connectivity provides an ever-growing social linguistic background in social media and internet networks. New language-games are emerging and old ones adopt or are coopted to offer semantics for the new. For instance, shopping is no longer a language-game between a customer and a salesperson in a store and filling a shopping cart is no longer putting objects in a supermarket container with wheels. Similarly, the identity, career, even life history of an individual leaves distinct footprints in cyberspace including language-games declaring credential, confessing beliefs, indicating preferences in modalities that include, but are not limited to, words, pictures, video, numbers, and community characteristics. AI that recognizes, constructs, and protects individual identity will construct meanings in a social linguistic background that it owes to possess but transcend explanation capabilities in order to judge and evaluate other AI entities. As illustrated in Fig. 1.3, connectivity becomes an infrastructure for community, a social linguistic background potentially as enriching for individual development in the lifespan as communities, which are not replaced but augmented and brought to the individual on demand.

Guideline 5: Transitivity. Comparisons and evaluations of AI can be made when specified conditions are met but the real test is transitivity in judgment. For instance, in Fig. 1.4 we see three types of interactions between machines. The first is an exchange. This is similar to a face-to-face meeting between individuals; usually it forms a strong basis for evaluating the relation or the potential for a relation especially at an early stage (e.g., love at first sight). The second is a direct evaluation based on the

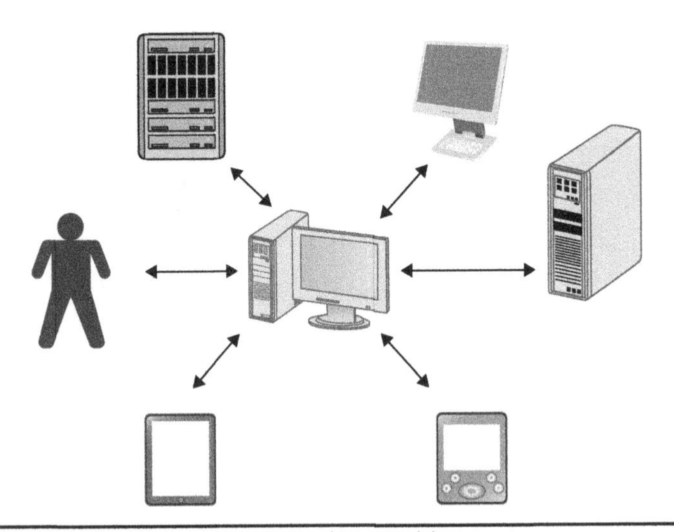

Figure 1.3 Social characteristics and connectivity.

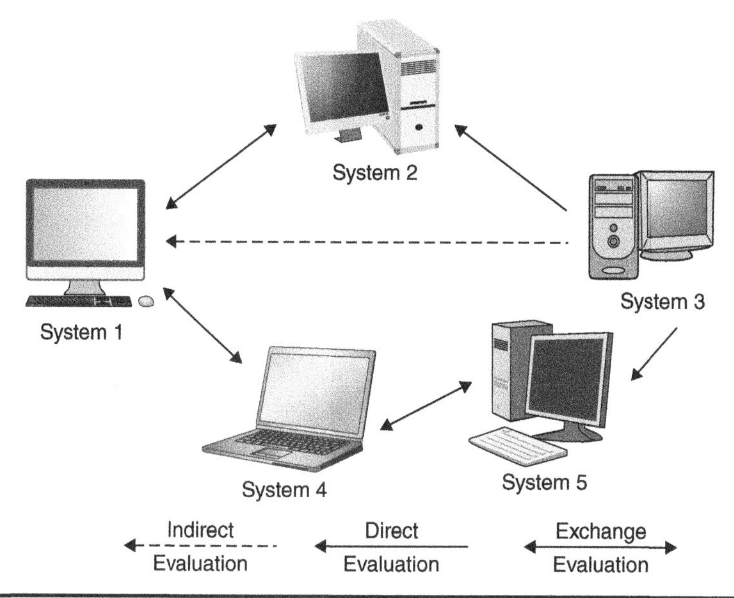

FIGURE 1.4 AI evaluation proceeds transitively from face-to-face (exchange) to friend-of-a-friend (direct) to six degrees of separation (indirect).

"friend of a friend" type approach of some direct interaction between third parties, which can also be the basis of an evaluation. Lastly, the third and more indirect basis of evaluation is akin to the so-called "degrees of separation" between two persons. In theory, humans including those that no longer exist physically are separated by each through six degrees of separation (that is, through six other humans). Transitivity in this sense means reaching out from the immediate to the distant through several stages. In all cases, categorizations of similarity or difference in the social linguistic background are essential and hence fuzziness is indispensable.

The evaluation of AI systems within a linguistic milieu using the above guidelines can proceed from human evaluators to machine evaluators. Humans can act as coaches, teachers, or mentors. But eventually machines such as *digital twins* should be able to judge on their own.[5] This is a dynamic process that involves growth stages and probably its own maturation or developmental characteristics. That being said, though, it is not a metaphysical problem. With the tools of science overall and the extraordinary flexibility fuzziness affords us, we are not far from where AI could address efficiently and over long time periods extending to decades or even centuries daunting problems in big data,

[5]*Digital twins* are examined in detail in Chap. 6.

energy security, climate risk mitigation, health, defense, and education infrastructures as well as entertainment and economics.

What our discussion has, so far, brought to the foreground is the idea that intelligence is not an innate property of an entity, be it human or machine. Rather, it is demonstrable via action in a social setting, and it is attributed as a property to its bearer if its success is socially acknowledged (this is in a broader, updated sense, the *imitation game* of Alan Turing). More than two millennia before Turing, Aristotle distinguished the *possession* of knowledge from the *use* of knowledge. The possession he called *episteme*, while the use he called *techne*; these terms roughly correspond to *science* and *engineering*, respectively, in modern parlance. Fuzzy logic becomes, thus, an enabler for transforming *episteme* (possessed knowledge) to *techne* (knowledge in use) and produce from it effective intelligence.[6]

References

Foerster, H. V., *Observing Systems*, 2nd edition, Intersystems Publications, Seaside, California, 1984.

Glasersfeld, E. V., *The Construction of Knowledge: Contributions to Conceptual Semantics*, Intersystems Publications, Seaside, California, 1987.

Manheim, J. H., *The Genesis of Point Set Topology*, Pergamon Press, Oxford, 1964.

Nguyen, H. T., Kosheleva, O. M., and Kreinovich, V., Is the Success of Fuzzy Logic Really Paradoxical?: Toward the Actual Logic Behind Expert Systems, *International Journal of Intelligent Systems*, Vol. 11, pp. 295–326, 1996.

Pitcher, G. (ed.), *Wittgenstein: The Philosophical Investigation*, Anchor Books, New York, 1966.

Tomasello, M., *A Natural History of Human Thinking*, Harvard University Press, Cambridge, Massachusetts; London, England, 2014.

Whitehead, A. N. and Russel, B., *Principia Mathematica*, Cambridge University Press, Cambridge, 1925.

Wittgenstein, L., *Philosophical Investigations*, 3rd edition, Pearson, 1953.

Wittgenstein, L., *Tractatus Logico-Philosophicus*, Harcourt, Brace & Company, New York, 1922.

Zadeh, L. A., Fuzzy Sets, *Information and Control*, Vol. 8, pp. 338–353, 1965.

Zadeh, L. A., Making Computers Think Like People, *IEEE Spectrum*, Vol. 21, Issue 8, pp. 26–34, August 1984.

[6]Aristotle also called for a third type of knowledge, *phronesis*, which roughly corresponds to *wisdom* and concerns the ethical application of knowledge, which we can find embedded in Turing's test (the imitation game is set up to "cheat" an observer into thinking that a machine "thinks," but do so while doing no harm).

CHAPTER 2
The Mathematics of Fuzzy Logic

This chapter gives an overview of the basic concepts and definitions of algebraic operations and fuzzy sets. Additionally, the one-dimensional and multidimensional aspects of fuzzy categorizations are thoroughly explained using if/then rules and linguistic descriptions. Practical worked examples are included to establish the basic terms and nomenclature.

2.1 Notating Fuzzy Sets

Fuzzy sets can model the words, sentences, and phrases we use. To compute with words, we need to familiarize ourselves with the mathematics of fuzzy sets, their notations, and operations. Different symbols and operations such as the maximum, *max*, or minimum, *min*, give a somewhat exotic appearance to fuzzy mathematics. They turn out to be easy and straightforward to use—*max* is like addition and *min* like multiplication.

We recall that a set A is comprised of several elements, which can be finite or infinite, but have a definite and fixed membership to the set. If an element belongs to the set, we say that its membership is 1 (one), and if not, it is 0 (zero). Usually, the elements of zero membership are not listed at all in the description of the set A. The elements of a set under consideration belong to a stable, unchangeable set, which is called the *universe of discourse*. The elements x of a universe of discourse X belong to the universe of discourse that is, $X = \{x\}$. This is analogous to a situation in the desert island experiment we encountered in Chap. 1. It takes time and effort to decide what belongs and what does not belong to A, and for this decision to bear fruits, it will involve acceptance from other members of the public (in the form of communication or mutual understanding) and, above all, produce results.

Mathematically, we can express whether an element x of a universe of discourse, X, belongs to the set A, through the characteristic function of the set A, that is,

$$\chi_A(x) = \begin{cases} 1 & \textit{iff} \quad x \in A \\ 0 & \textit{iff} \quad x \notin A \end{cases} \tag{2.1}$$

which can also write as

$$\chi_A(x) : X \rightarrow \{0, 1\}, \qquad \forall\, x \in X \tag{2.2}$$

In the notation of Eqs. (2.1) and (2.2), $\chi_A(x)$ is the characteristic function of the set A, that is, a function that unequivocally defines its identity, and it does so by taking every element x of the universe of discourse X, asking the question "does x belong to A?" If the answer is Yes, the characteristic function outputs the value 1; if it is No, it outputs 0. Additionally, *iff* is a shorthand for "*if and only if*," a strict mathematical condition, $\forall\, x \in X$ means "for every element x that belongs to the X, and $x \in A$" means "x belongs to A" while $x \notin A$ means "x does not belong to A." Thus, Eq. (2.2) should be read as follows: "*The characteristic function of A, denoted as* $\chi_A(x)$, *is such (that's the meaning of ':') that it maps (that's the meaning of the arrow '→') the entire universe of discourse X to the valuation set (meaning the set of labels) of the pair of numbers 0 and 1, denoted as* $\{0, 1\}$, *for every single element x that belongs to X.*"

Mathematically, fuzziness is introduced when the characteristic function in Eqs. (2.1) and (2.2) is generalized by changing the pair $\{0, 1\}$ (which has only two numbers) to the interval from 0 to 1, denoted as $[0, 1]$, which contains infinite numbers. It may seem counterintuitive, but mathematically speaking, in the interval $[0, 1]$, there are as many numbers as in the interval from negative infinity to positive infinity, that is, the interval $[-\infty, +\infty]$. Fuzziness is about what Noam Chomsky calls "discrete infinity," a term that refers to a phenomenon of language where we have a finite set of words, each capable of holding an infinity of meanings. The first step in mathematically modeling such an assertion is to generalize the notion of membership by changing the valuation set from the pair $\{0, 1\}$ to the interval from 0 to 1, or $[0, 1]$, which contains infinite numbers.

A fuzzy set A on the universe of discourse X is defined by a generalized characteristic function, from now on called *membership function* and denoted as $\mu_A(x)$, as follows:

$$\mu_A(x) : X \rightarrow [0, 1], \qquad \forall\, x \in X \tag{2.3}$$

where, $\mu_A(x)$ is the membership function of the set A, and x is an element of the universe of discourse X. Equation (2.3) is saying that for every element of x that belongs to the universe of discourse X, denoted as $x \in X$, we assign a number between 0 and 1 that uniquely indicates its degree of membership to the set A.

Given this situation, the minimum information that we must have about an element of A is not only the name of the element, denoted by x, but also the degree of its membership to the set A, denoted by $\mu_A(x)$. The pair $\mu_A(x)$ and x must always be together; hence, we call it "a singleton" and we denote it as $\mu_A(x)/x$. The singleton is the atomic minimum that must be said about an element of a set. It should be noted that the slash in the singleton notation is just a convenient marker, it does not signify division. In this way, the fuzzy set A is considered as the set of singletons $\mu_A(x)/x$ and can be written in set notation as

$$A = \{\mu_A(x)/x\}$$

An alternative notation, used more often in the literature explicitly indicates a fuzzy set as the *union* of all $\mu_A(x)/x$ singletons, that is,

$$A = \sum_{x_i \in X} \mu_A(x_i)/x_i \text{ or} \\ A = \int_X \mu_A(x)/x \tag{2.4}$$

The *summation* and integral signs in Eq. (2.4) indicate the *union* of all singletons (the union operation in set theory is like "addition") for a discrete and continuous universe of discourse, respectively.

The *support* of a fuzzy set A is the set of all those elements of the universe of discourse X that have nontrivial membership, that is, membership greater than zero, or

$$\mu_A(x) > 0$$

An important concept in fuzzy logic is that of a *fuzzy variable*, which is a variable that takes on fuzzy sets as its values. For instance, the *temperature* of an environment may be considered a fuzzy variable taking on a set of values, which we linguistically describe as *LOW*, *MEDIUM*, and *HIGH*, but for each one of these words there is a fuzzy set with its own membership function defined over the universe of discourse of numerical temperatures as could have been measured by a thermometer placed in that particular environment. These words, which we call "fuzzy values," are precisely defined through fuzzy sets. Again, underneath each one of them there is a fuzzy set defined by a membership function such as in Eq. (2.4).

More generally, the values of a fuzzy variable don't need to be single words; they can also be phrases in some predefined language, e.g., a combination of fuzzy values joined together with logical connectives such as *AND*, *OR*, *NOT*, and linguistic hedges. In such situations, fuzzy variable can be used to construct linguistic relations in the form *if/then* rules comprised of simple propositions in association. For example, the values of the fuzzy variable *temperature* may be expressed as *HIGH, NOT HIGH, RELATIVELY HIGH, VERY HIGH, NOT VERY HIGH, VERY MUCH HIGH, HIGH BUT NOT VERY HIGH, SUFFICIENTLY HIGH*. Such values are compound words comprised of the word *HIGH*, the negation *NOT*, the connectives *AND* and *BUT*, and the hedges *VERY, RELATIVE, ENOUGH, ALMOST*.

The fuzzy variable *temperature* becomes what is also known as a *linguistic variable* whose arguments are words modeled by fuzzy sets. If at a given instance, a linguistic variable takes a specific value, the resulting structure is called a proposition. It typically has the form

$$p^1: \quad x \text{ is } A \tag{2.5}$$

In Eq. (2.5), the proposition p^1 is the minimum that can be said about anything, literally a minimal statement which may be meaningful within a given language-game.

The conditional dependence of one linguistic variable upon another (independent of the first linguistic variable) forms a fuzzy *if/then* rule, often denoted by R. The rule may be written as:

$$R: \text{ if } \quad p^1 \quad \text{then} \quad p^2 \tag{2.6}$$

where p^1 and p^2 in Eq. (2.6) are fuzzy propositions instantiating (means taking on specific values) the associate fuzzy variables, which have the form:

$$\begin{aligned} p^1&: \quad x \text{ is } A \\ p^2&: \quad y \text{ is } B \end{aligned} \tag{2.7}$$

Combining Eqs. (2.6) and (2.7), we can write the *if/then* rule R as:

$$R: \quad \text{if } x \text{ is } A \quad \text{then } y \text{ is } B \tag{2.8}$$

where A is a fuzzy set of the universe of discourse X, and B is a fuzzy set on the universe of discourse Y, while elements x and y belong to X and Y, respectively (or as we shall see a little later to the Cartesian product $X \times Y$). If it is clear what is the universe of discourse for A and B, we sometimes write Eq. (2.8) in a shorthand as:

$$R: \quad A \quad \rightarrow \quad B \tag{2.9}$$

The simple representation of a rule in Eq. (2.9) facilitates operations with aggregations of rules, which we call *Rule Clusters* (RC) as we shall see in Chaps. 3 and 4.

To the fuzzy values A and B in Eqs. (2.8) and (2.9), we can attribute words that define their values. For example,

$$\textit{if} \;\; x \;\; \textit{is} \;\; \textit{SMALL} \;\; \textit{then} \;\; y \;\; \textit{is} \;\; \textit{VERY} \;\; \textit{LARGE}$$

or,

$$if\ error\ is\ NL\ then\ output\ is\ PB$$

where *NL* stands for *Negative Small* (that is, small on the negative side of a universe of discourse) and *PB* stands for *Positive Big* (big on the positive side of a universe of discourse). Two or more fuzzy *if/then* rules may be combined (one inside the other) so that they can form an embedded fuzzy statement of the form:

$$R:\ if\ \ p^1\ \ then\ \ (if\ \ p^2\ \ then\ \ p^3) \tag{2.10}$$

It is easy to see that the compound rule of Eq. (2.10) is made out of two simple *if/then* rules, that is,

$$\begin{aligned} R^1 &: if\ \ p^1\ \ then\ \ R^2 \\ R^2 &: if\ \ p^2\ \ then\ \ p^3 \end{aligned} \tag{2.11}$$

where the first rule R^1 in Eq. (2.11) uses the second rule R^2 as a consequence, that is, as action to be taken if proposition p^1 on the *Left Hand Side* (*LHS*) of R^1 is activated, that is, if it is deemed correct to some degree. This kind of computations parameter is called the *Degree of Fulfillment* (*DOF*) of the rule and provides a useful parameter for application of deep learning approaches in AI (a process to which we will devote considerable attention later on under the subjects of *inferencing* and *composition*). This is done by placing the second rule R^2 on the *Right Hand Side* (*RHS*) of the first. For example, such a situation can be seen in a compound rule found in control applications, that is,

$$if\ error\ is\ NME\ then\ (if\ \ change-in-error\ \ is\ \ PME\ then\ output\ \ is\ \ PME)$$

where *error* is a fuzzy variable taking the value *Negative Medium* (*NME*) in the *LHS* of the rule, while on the *RHS* we have an entire rule having change-in-error in its own *LHS* taking the value *Positive Medium* (*PM*) and in its own *RHS* the fuzzy variable *output* taking the value *Positive Medium* (*PME*).

Two or more fuzzy conditional statements can be combined with the connective *ELSE* in order to form a fuzzy algorithm R^N of the form:

$$R^N:\ \ if\ R^1\ ELSE\ R^2\ ELSE\ R^3 \ldots\ ELSE\ R^n$$

The subject of fuzzy algorithms or *Rule Clusters* (*RC*) is very important for all kinds of applications where the coordination of activities requires a language-game to be formed. We will expand on how to use them in Chaps. 3 and 4.

2.2 Fuzzy Set Operations

In fuzzy logic, the operators *min* and *max* have special significance. Learning how to use them is easy, although at first they look somewhat unfamiliar. In this section, we start with their definition and proceed with examples of their use to build familiarity.

The *min* and *max* operators can be used between two elements (binary) or applied to a list of elements depending on the context. They can operate on sets or matrices and on each element separately, as well as on independent elements. In all cases, they identify and select out the smallest or the largest element. The *min* and *max* of two elements a and b are defined as follows:

$$a \wedge b = \min(a, b) = \begin{cases} a & if \quad a \le b \\ b & if \quad a > b \end{cases}$$
$$a \vee b = \max(a, b) = \begin{cases} a & if \quad a \ge b \\ b & if \quad a < b \end{cases} \tag{2.12}$$

The *min* and *max* of two fuzzy sets A and B (defined over the same universe of discourse) produce a set C by comparing each element $a \in A$, $b \in B$ and taking the smallest or largest value respectively as the element of C, that is,

$$\begin{aligned} C &= A \wedge B = \{\min(a, b)\}, \quad \forall\ a \in A, b \in B \\ C &= A \vee B = \{\max(a, b)\}, \quad \forall\ a \in A, b \in B \end{aligned} \tag{2.13}$$

When the operators are used as unitary operators, they are selecting the minimum (also called *infimum*) or the maximum (also called *supremum*) element from all the elements of a set, for example,

$$\begin{aligned} a &= \wedge A = inf\ A, \quad a \in A \\ a &= \vee A = sup\ A, \quad a \in A \end{aligned} \tag{2.14}$$

The operations shown in Eqs. (2.13) and (2.14) can be applied to lists of elements acting as functions that pick up the smallest or largest element in a list, e.g., for the min operator we have,

$$\begin{aligned} a &= \bigwedge(a_1, a_2, a_3, \ldots, a_m) \\ &= a_1 \wedge a_2 \wedge a_3 \wedge \ldots \wedge a_m \\ &= \bigwedge_{k=1}^{m}(a_k) \\ A &= \bigwedge(A_1, A_2, A_3, \ldots, A_m) \\ &= A_1 \wedge A_2 \wedge A_3 \wedge \ldots \wedge A_m \\ &= \bigwedge_{k=1}^{m}(A_k) \end{aligned} \tag{2.15}$$

In situations where the elements of a set are functions of one variable, the operators can be expressed in the case of min as:

$$a = \bigwedge_x (a(x)),\ x \in X \tag{2.16}$$

Similar uses hold in Eqs. (2.15) and (2.16) if we were to replace the *min* ($\wedge$) operator with the *max* ($\vee$) operator. Finally, it should be noted that in operations where we use *min* and *max*, we use the same algebraic rules that we would use in multiplication and addition correspondingly.

2.3 Operations with Fuzzy Sets

We turn our attention to fuzzy set operations keeping in mind that we use one-dimensional fuzzy sets on the same universe of discourse (Tsoukalas, 1997). The following terms and operations are useful in the broader task referred to as the *symbolization of reality* in Chap. 1.

2.3.1 Empty Fuzzy Set

A fuzzy set A is called *empty* if and only if (a strict mathematical requirement that we have seen in Sec. 2.1 is denoted by the shorthand *iff*) its membership function is zero everywhere in the universe of discourse X, that is,

$$A = \varnothing \ \mathit{iff}\ \ \mu_A(x) = 0,\ \forall\ x \in X \tag{2.17}$$

2.3.2 Normal Fuzzy Set

A fuzzy set is called *normal* if there is at least one element x_o in the universe of discourse that has full membership to the set, that is, where the membership function equals 1, i.e.,

$$\mu_A(x_o) = 1 \tag{2.18}$$

In Eq. (2.28), full membership pertains to one element. But this is not necessary in fuzzy logic; there can be more than one element in the universe of discourse that satisfy normalcy.[1]

2.3.3 Equality of Fuzzy Sets

Two fuzzy sets A and B are said to be *equal* if their membership functions are equal throughout the universe of discourse, i.e.,

$$A \equiv B,\quad \mathit{if}\quad \mu_A(x) = \mu_B(x) \tag{2.19}$$

[1]It should be noted that the term normal does not refer to the area under the curve of the membership function. It simply means what the definition says: at least one point, maybe more, have full membership to the set.

2.3.4 Union of Fuzzy Sets

The *union* of two fuzzy sets A and B, each defined over the same universe of discourse X, is a new fuzzy set, denoted as $A \cup B$, also on the universe of discourse X, with membership function which is the maximum of the grades of membership of every x to A and B, i.e.,

$$\mu_{A \cup B}(x) \equiv \mu_A(x) \vee \mu_B(x) \tag{2.20}$$

The *union* of two fuzzy sets is related to the logical operation of *disjunction* (*OR*). Equation (2.20) can be generalized to any number of fuzzy sets over the same universe of discourse.

2.3.5 Intersection of Fuzzy Sets

The *intersection* of two fuzzy sets A and B is a new fuzzy set, denoted as $A \cap B$ and defined over the same universe of discourse X. The new set has membership function obtained by forming the minimum of the grades of membership of every x to the sets A and B, i.e.,

$$\mu_{A \cap B}(x) \equiv \mu_A(x) \wedge \mu_B(x) \tag{2.21}$$

The *intersection* of two fuzzy sets is related to *conjunction* (*AND*) in fuzzy logic. The definition of *intersection* in (2.21) can be generalized to any number of fuzzy sets over the same universe of discourse.

2.3.6 Complement of a Fuzzy Set

The *complement* of a fuzzy set A is a new fuzzy set, $\bar{A}$, with membership function:

$$\mu_{\bar{A}}(x) \equiv 1 - \mu_A(x) \tag{2.22}$$

Fuzzy set *complementation* is equivalent to *negation* (*NOT*) in fuzzy logic.

2.3.7 Product of Two Fuzzy Sets

The product of two fuzzy sets A and B, defined on the same universe of discourse X, is a new fuzzy set $A \cdot B$, with membership function that equals the algebraic product of the membership functions of A and B,

$$\mu_{A \cdot B}(x) \equiv \mu_A(x) \cdot \mu_B(x) \tag{2.23}$$

The product of two fuzzy sets can be generalized to any number of fuzzy sets on the same universe of discourse.

2.3.8 Multiplying a Fuzzy Set by a Crisp Number

We can multiply the membership function of a fuzzy set A by the number a to obtain a new fuzzy set called *product* $a \cdot A$. Its membership function is:

$$\mu_{aA}(x) \equiv a \cdot \mu_A(x) \tag{2.24}$$

The operations of multiplication and raising a fuzzy set to a power that we see next are useful for modifying the meaning of linguistic terms (Zadeh, 1975).

2.3.9 Power of a Fuzzy Set

We can raise fuzzy set A to a power α (positive real number) by raising its membership function to α. The *α power of A* is a new fuzzy set, A^α with membership function:

$$\mu_{A^\alpha}(x) \equiv [\mu_A(x)]^\alpha \tag{2.25}$$

When $\alpha = 2$, that is, when a fuzzy set A is raised to the second power, this is equivalent to linguistically modifying A through the modifier *VERY* (Zadeh, 1983). Thus the square of the membership function of A = {*TALL PEOPLE*} represents the fuzzy set A^2 = {*VERY TALL PEOPLE*}.

Raising a fuzzy set to the second power is a particularly useful operation and therefore has its own name. It is called *concentration* or *CON*. Taking the square root of a fuzzy set is called *dilation* or *DIL* (an operation useful for representing analytically the linguistic modifier *MORE OR LESS*).

2.4 Alpha Cuts and Resolution

With any fuzzy set A we can associate a collection of crisp sets known as *α-cuts* (*alpha-cuts*). An *α-cut* is a crisp set consisting of elements of A which belong to the fuzzy set at least to a degree α. They offer a method for resolving any fuzzy set in terms of constituent crisp sets (something analogous to resolving a vector into its components). It should be noted that *α-cuts* are crisp, *not* fuzzy sets. Formally, the *α-cut* of a fuzzy set A, denoted as A_α, is the crisp set comprised of all the elements x of a universe of discourse X for which the membership function of A is *greater than or equal* to α, i.e.,

$$A_\alpha = \{x \in X \mid \mu_A(x) \geq \alpha\} \tag{2.26}$$

where α is a parameter in the range $0 < \alpha \leq 1$; the vertical bar "|" in Eq. (2.26) is a shorthand for "such that."

Consider, for example, the fuzzy set A of *small integers* given by:

$$A = 1.0/1 + 1.0/2 + 0.75/3 + 0.5/4 + 0.3/5 + 0.3/6 + 0.1/7 + 0.1/8$$

The *0.5-cut* of A is simply the crisp set $A_{0.5} = \{1, 2, 3, 4\}$.

It can be shown that *α-cuts* provide a useful way both for resolving a membership function in terms of constituent crisp sets as well as for synthesizing a membership function out of crisp sets, a process known as the *resolution principle* (Tsoukalas, 1997).

Example 2.1 Union, Intersection, and Complement of Fuzzy Sets Consider two fuzzy sets A and B defined over the universe of discourse $X = \mathbb{R}^+$ (positive real numbers) by the membership functions

$$\mu_A(x) = \frac{1}{1+[3(x-1.6)]^2}$$
$$\mu_B(x) = \frac{1}{1+[1(x-3.2)]^2} \tag{E2.1A}$$

The membership functions of A and B are shown in Fig. 2.1. Such graphical representations of the membership functions are very useful in fuzzy logic and are given a special name. They are called Zadeh diagrams and can be thought of as analogous in importance to the Venn diagrams employed in crisp sets. Fuzzy set A may be thought of as describing the neighborhood of numbers which we could linguistically describe by the statement "*about 1.6*" and fuzzy set B as describing the neighborhood "*about 3.2.*" For plotting the Zadeh diagrams, we limit the universe of discourse to numbers between "0" and "'10," and would like to find the *union* and *intersection* of A and B and the *complement* of B.

The union of A and B is the new set $A \cup B$, which has membership function constructed by taking the maximum grade of membership of each element x of the universe of discourse to either A or B in accordance with Eq. (2.20). Figure 2.1 shows the membership function of the *union* $A \cup B$. The linguistic description of $A \cup B$ is "*about 1.6 OR 3.2.*"

Similarly the membership function of the *intersection* of fuzzy sets A and B, shown in Fig. 2.2 represents the new fuzzy set "*about 1.6 OR 3.2.*" We observe that although the union of A and B is a *normal* fuzzy set, the intersection shown in Fig. 2.2 is not, because fuzzy set $A \cap B$ has no point in the universe of discourse with a membership grade of "1."

Finally, the complement of fuzzy set A is a new fuzzy set with membership function given by Eq. (2.22). Figure 2.3 shows the membership function $\mu_{\bar{A}}(x)$

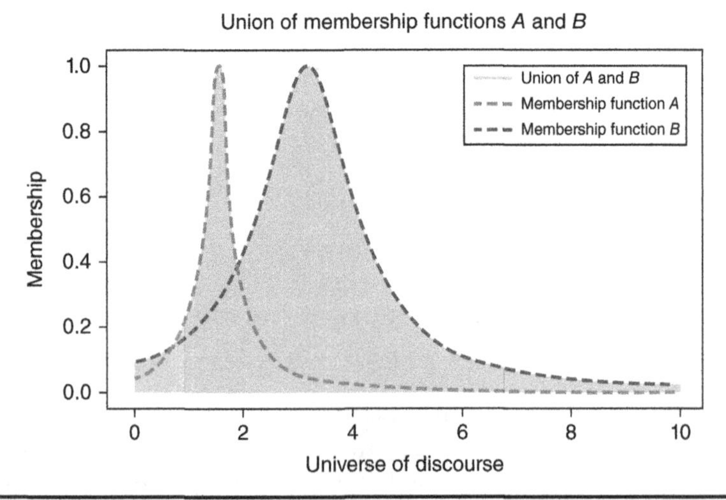

FIGURE 2.1 Zadeh diagrams for the union of sets A and B which have membership functions given in Eq. (E2.1A).

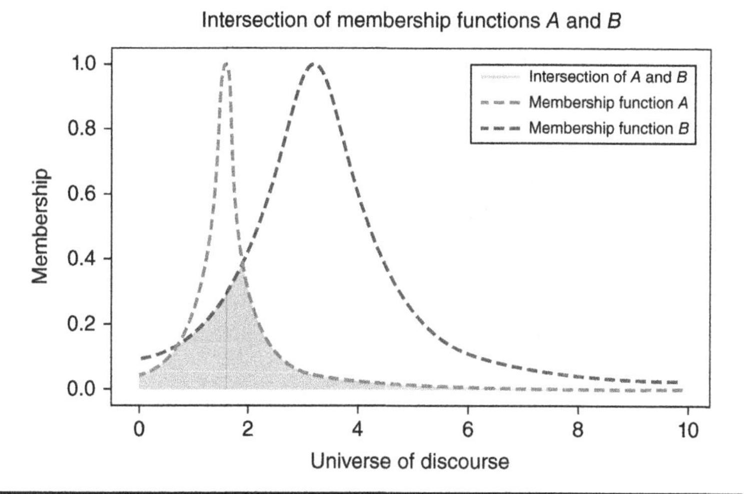

Figure 2.2 Zadeh diagrams for the intersection of sets A and B.

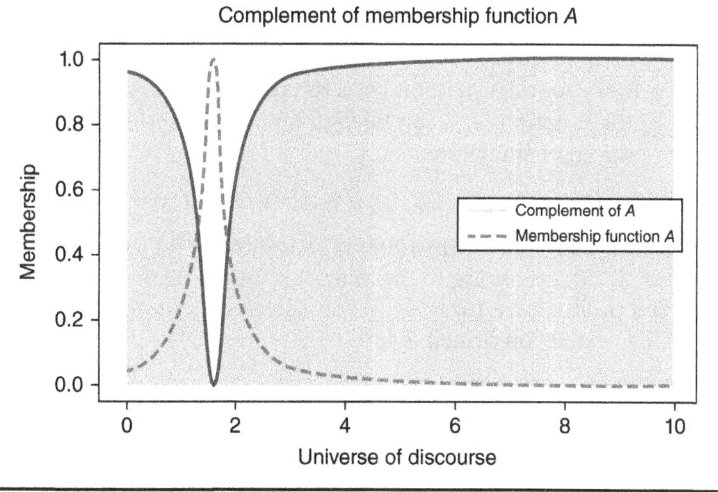

Figure 2.3 Zadeh diagrams for the union of sets A and B.

of the *complement* of A, that is $\bar{A}$. The complement $\bar{A}$ represents the logical negation (*NOT*) of A, that is, the set "*NOT about 3.6.*" ◊[2]

2.5 The Extension Principle

While fuzzification operations such as the ones we saw in Sec. 2.3 are useful for fuzzifying individual sets or singletons, more general mathematical expressions may also be fuzzified when the quantities

[2]The end of an example is indicated by the marker "◊."

involved are fuzzified. For example, the output of mathematical operations when the inputs are fuzzy sets becomes fuzzy as well. In other words, when we fuzzify a domain, and there is a mapping from that domain to a range, the range becomes fuzzified too. The *extension principle* is a mathematical tool for extending crisp mathematical notions and operations to the milieu of fuzziness. It provides the theoretical warranty that fuzzifying the parameters or arguments of a function results in computable fuzzy sets. It is an important principle, and we will use it on several occasions, particularly in conjunction with fuzzy relations the principle model of our language-games. We offer next a description of the extension principle (Zadeh, 1975; Dubois, 1980).

Suppose that we have a function f that maps elements $x_1, x_2, \ldots, x_n$ of a universe of discourse X to another universe of discourse Y, i.e.,

$$\begin{aligned} y_1 &= f(x_1), \\ y_2 &= f(x_2), \\ &\cdots \\ y_n &= f(x_n), \end{aligned} \tag{2.27}$$

If a fuzzy set A is defined over the elements $x_1, x_2, x_3, \ldots, x_n$ (the input to the function f), A can be mathematically described as a collection (union) of singletons,

$$A = \mu_A(x_1)/x_1 + \mu_A(x_2)/x_2 + \cdots + \mu_A(x_n)/x_n \tag{2.28}$$

If the domain of our function f becomes fuzzy, for example the set A of Eq. (2.28), according to the extension principle the range is also fuzzified producing a fuzzy set B that can be computed by inputting A to f. The extension principle tells us that the fuzzy set B is given by

$$B = f(A) = \mu_A(x_1)/f(x_1) + \mu_A(x_2)/f(x_2) + \cdots + \mu_A(x_n)/f(x_n) \tag{2.29}$$

where every single image of x_i under f, i.e., $y_i = f(x_i)$, becomes fuzzy to a degree of $\mu_A(x_i)$. Since functions are *many-to-one* mappings, it is conceivable that several x's may map to the same y. For a certain y_0 we may have more than one x, let us say both x_2 and x_{13} in Eq. (2.27) are mapping to y_0. Hence, we have to decide which of the two membership values, $\mu_A(x_2)$ or $\mu_A(x_{13})$, we should take as the membership value of y_0. The extension principle says that the *maximum* of the membership values of these elements in the fuzzy set A ought to be chosen as the grade of membership of y_0 to the set B, that is,

$$\mu_B(y_0) = \mu_A(x_2) \vee \mu_A(x_{13}) \tag{2.30}$$

If on the other hand no element x in X is mapped to y_0, that is to say no inverse image of y_0 exists, then the membership value of the set B at y_0 is zero. Having accounted for these two special cases (many x's mapping to the same y and no inverse image for a certain y), we can compute the set B, i.e., the grades of membership of elements y in Y produced by the mapping $f(A)$, using Eq. (2.29).

In many applications, the interpretation of numerical data may not be precisely known. We consider this type of data to be fuzzy. Using the extension principle, it is possible to adapt ordinary algorithms used with precise data, to the case where the data is fuzzy. Example 2.2 offers a mathematical illustration of the extension principle.

Example 2.2 Using the Extension Principle to Fuzzify a Straight Line As an illustration of how the extension principle may be used, consider a function f that maps points from the x-axis to y-axis in the Cartesian plane according to the equation of a line $\mathcal{L}$ in the coordinate plane that is:

$$Ax + By + C = 0 \tag{E2.2A}$$

where the coefficients A and B in (2.32) must always satisfy $A^2 + B^2 \neq 0$.

We let the points of line $\mathcal{L}$ to be mapped from any combination of $S(x_1, y_1)$ and will find the fuzzy set A defined on $x_1, x_2, \ldots, x_n$ of line $\mathcal{L}$ given as

$$y = f(x) = -x + 10 \tag{E2.2B}$$

For the fuzzy set A (not related to the coefficient of the general form of a line given above) we will be using a triangular shape for its membership function.

It is clear from the given line that the coefficients are $A = 1$, $B = 1$, and $C = -10$. The fuzzy set A defined on $x_1, x_2, \ldots, x_n$ is known to be:

$$A = \mu_A(x_1)/x_1 + \mu_A(x_2)/x_2 + \cdots + \mu_A(x_n)/x_n$$

where μ_A is the membership function, which describes the degree of fuzziness. The membership function using the triangular shape is generally modeled as:

$$\mu(x) = \begin{cases} 0, & x \leq a \\ \left(\dfrac{x-a}{b-a}\right), & a \leq x \leq b \\ \left(\dfrac{c-x}{c-b}\right), & b \leq x \leq c \\ 0, & x \geq c \end{cases} \tag{E2.2C}$$

Choosing $a = 2$, $b = 4$, and $c = 10$ in Eq. (E2.2C), we obtain the following membership function:

$$\mu_A(x) = \begin{cases} 0, & x \leq 2 \\ \left(\dfrac{x-2}{2}\right), & 2 \leq x \leq 4 \\ \left(\dfrac{10-x}{6}\right), & 4 \leq x \leq 10 \\ 0, & x \geq 10 \end{cases} \tag{E2.2D}$$

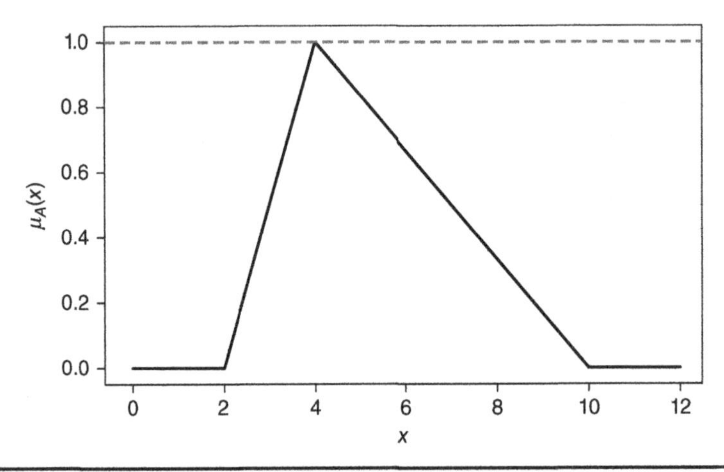

FIGURE 2.4 Zadeh diagram for the triangular shaped fuzzy set *A*.

In Fig. 2.4, we plot Eq. (E2.2D) to produce the Zadeh diagram of the set *A*. Thus, the Zadeh diagram is comprised of two lines, one for the *Left Side* (*LS*) and another for the *Right Side (RS)* of the triangle, and the set *A* can be written as the union of singletons for LS and RS, that is,

$$A = \int_{2 \le x \le 4} \left(\frac{x-2}{2}\right) / x + \int_{4 \le x \le 10} \left(\frac{10-x}{6}\right) / x \qquad \text{(E2.2E)}$$

In order to find the fuzzy set *B* induced by the fuzzified inputs $f(x_1), f(x_2), \dots, f(x_n)$ of line $\mathcal{L}$ we use the "Extension Principle" to calculate the output of a function *f*. Thus, the fuzzy set *B* output is obtained in the form of:

$$B = \mu_A(x_1)/f(x_1) + \mu_A(x_2)/f(x_2) + \cdots + \mu_A(x_n)/f(x_n)$$

Having the *x* values fuzzified by the fuzzy set *A*, we want to know the effect of fuzzification on *y*. First, we solve for *x* in the given equation. Thus, we obtain:

$$y = -x + 10 \rightarrow x = -y + 10$$

Then, we substitute *x* so we get:

$$B = \int_{0 \le y \le 6} \left(\frac{y}{6}\right) / y + \int_{6 \le y \le 8} \left(\frac{8-y}{2}\right) / y$$

where the membership function of the fuzzy set *B* is shown in Fig. 2.5 and is obtained as:

$$\mu_B(y) = \begin{cases} 0, & y \le 0 \\ \left(\frac{y}{6}\right), & 0 \le y \le 6 \\ \left(\frac{8-y}{2}\right), & 6 \le y \le 8 \\ 0, & y \ge 8 \end{cases}$$

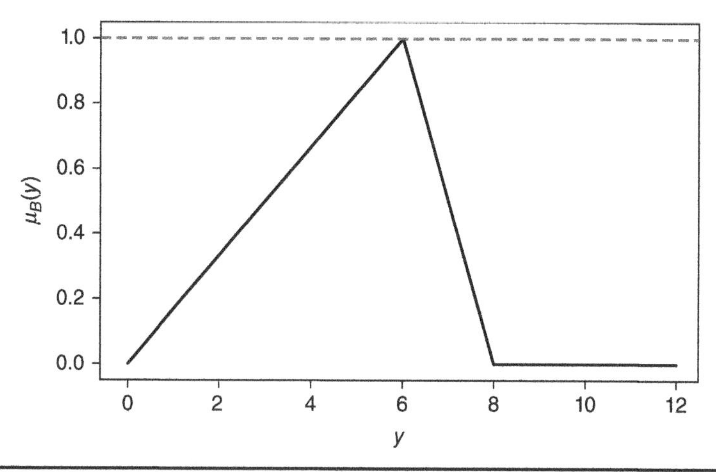

Figure 2.5 Zadeh diagram for the set B induced by A and the mapping of a straight line.

Example 2.3 Fuzzification of Lines Using Bell-Shaped Membership Functions If in Example 2.2 we use bell-shaped membership function instead of a triangle, the fuzziness induced by a straight-line mapping is taking another form. In general, a bell-shaped function is given by

$$\mu(x) = \frac{1}{1 + \left|\frac{x-c}{a}\right|^{2b}} \tag{E2.3A}$$

where the parameters a, b, and c in Eq. (E2.3A) are used to tune the shape of the function, that is, to control the width, slope, and center, respectively. Choosing $a = 1$, $b = 1$, and $c = 6$, we obtain the following membership function for A:

$$\mu_A(x) = \frac{1}{1 + |x-6|^2} \tag{E2.3B}$$

The Zadeh diagram for A, that is, the graph of (E2.3B), is shown in Fig. 2.6.

The fuzzy set A as the union of singletons is analytically described by:

$$A = \int_{1 \le x \le 11} \left(\frac{1}{1 + |x-6|^2} \right) / x \tag{E2.3C}$$

The fuzzy set B induced by fuzzifying the inputs $f(x_1), f(x_2), \ldots, f(x_n)$ of line $\mathcal{L}$ is found by using the *Extension Principle*. The fuzzy set B, which is the output of the mapping, is obtained by passing the fuzziness through, i.e.,

$$B = \mu_A(x_1)/f(x_1) + \mu_A(x_2)/f(x_2) + \cdots + \mu_A(x_n)/f(x_n) \tag{E2.3D}$$

Having the x values fuzzified by the fuzzy set A, we want to know the effect of fuzzification on y as shown in Eq. (E2.3D). First, we solve for x in the given equation to obtain:

$$y = -x + 10 \rightarrow x = -y + 10$$

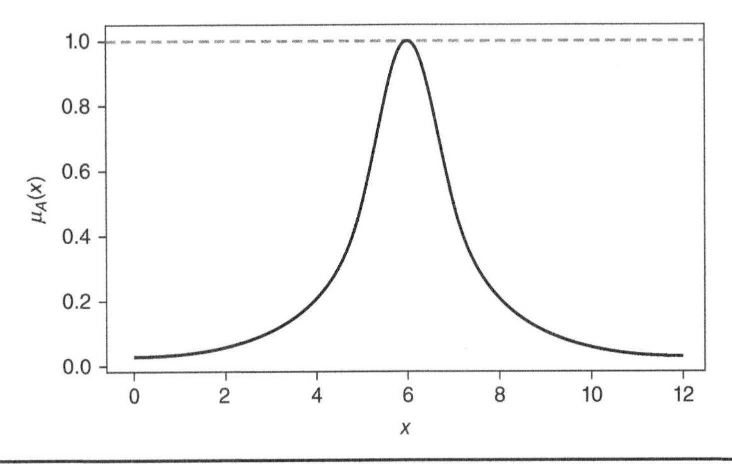

FIGURE 2.6 Zadeh diagram for a bell-shaped fuzzy set *A*.

Then, we substitute x for y to get:

$$B = \int_{-1 \le y \le 9} \left(\left(\frac{1}{1 + |y-4|^2} \right) \right) / y$$

Hence the membership function of the induced fuzzy set B is obtained as:

$$\mu_B(y) = \frac{1}{1 + |y-4|^2} \tag{E2.3E}$$

The Zadeh diagram of the induced fuzziness through the extension principle is shown in Fig. 2.7, which is a graph of Eq. (E2.3E). It should be noted that the bell shape is maintained because of the linear mapping, but its center has shifted to a different value on the y-axis.

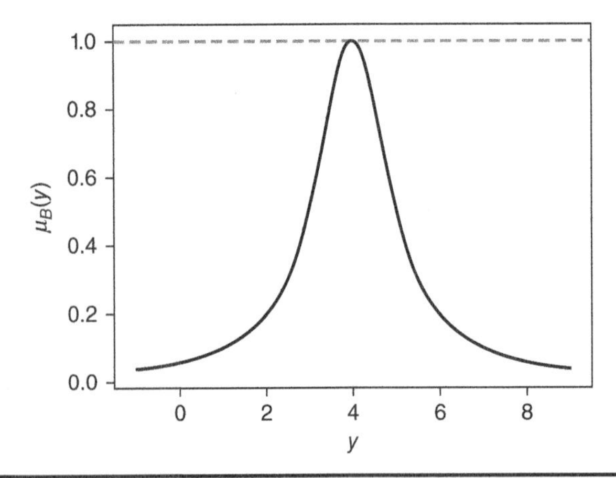

FIGURE 2.7 Zadeh diagram for the induced set B in Example 2.3.

References

Dubois, D., and Prade, H., *Fuzzy Sets and Systems: Theory and Applications*, Academic Press, Boston, 1980.

Dubois, D., and Prade, H., *Possibility Theory*, Plenum Press, New York, 1988.

Terano, T., Asai, K., and Sugeno, M., *Fuzzy Systems Theory and Its Applications*, Academic Press, Boston, 1992.

Tsoukalas, L. H., and Uhrig, R. E., *Fuzzy and Neural Approaches in Engineering*, Wiley, New York, 1997.

Zadeh, L. A., "Fuzzy Sets," *Information and Control*, Vol. 8, pp. 338–353, 1965.

Zadeh, L. A., "Probability Measure of Fuzzy Events," *Journal of Mathematical Analysis and Applications*, Vol. 23, pp. 421–427, 1968.

Zadeh, L. A., "Outline of a New Approach to the Analysis of Complex Systems and Decision Processes," *IEEE Transactions on Systems, Man and Cybernetics*, SMC-3, pp. 28–44, 1973.

Zadeh, L. A., "The Concept of a Linguistic Variable and Its Application to Approximate Reasoning," *Information Sciences*, Vol. 8, pp. 199–249, 1975.

Zadeh, L. A., "Fuzzy Sets as a Basis for Theory of Possibility," *Fuzzy Sets and Systems*, Vol. 1, pp. 3–28, 1978.

Zadeh, L. A., "A Computational Approach to Fuzzy Quantifiers in Natural Languages," *Computer and Mathematics*, Vol. 9, pp. 149–184, 1983.

Zadeh, L. A., "Fuzzy Logic," *IEEE Computer*, pp. 83–93, April 1988.

CHAPTER 3

Many-to-Many Mappings in AI and ML

Where terms and operations for linguistic descriptions are presented. The one-dimensional and multidimensional aspects of fuzzy categorizations are explained using if/then rules to compute with language.

3.1 Fuzzy Propositions and if/then Rules

In logic, most of the time we create models using *if/then* rules. Rules provide a convenient structure to encode what Wittgenstein calls "language-games." We recall from Chap. 1 that language is comprised of language-games, that is acts of coordination evaluated (that is, made meaningful) through and against a social linguistic background that provides the criteria for the correct use of words. AI can use these criteria to become actionable, corrective, and interpretable. In the same way that language cannot be wholly determined by private experience, AI does not reside in computers alone. It is found in the interactions of AI with humans or with computers—like language, AI **is an essentially social phenomenon**. In this chapter we will illustrate the practical side of these ideas and look at how fuzzy logic can bridge AI with language, that is the rapidly growing space of symbols and meanings that surround our increasingly digitalized lives.

In formulations of logic, a rule is comprised of one or more propositions on the left-hand side (*LHS*), associated with one or more propositions on the right-hand side (*RHS*). The syntax of these statements is familiar to all, and the statements closely resemble simple everyday life ***language-games***, including, but not limited to, prescriptions for *action* and *cause and effect* statements.

For instance, consider a typical morning phrase on a day that the weather forecast calls for rain, something like, "today it's going to be a rainy day, and therefore I need to have an umbrella." The phrase can easily be casted in the structure of a fuzzy *if/then* rule,

$$if\ (weather\ is\ RAINY)\quad then\quad (umbrella\ is\ NEEDED)$$

The *LHS* of this rule is comprised of a single elementary proposition having one linguistic variable called *weather* which is assigned the linguistic value *RAINY*. Similarly, the *RHS* has an elementary proposition with the linguistic value *umbrella* assigned the value *NEEDED*.

If an AI app is looking at umbrella descriptions in the social media space, then a higher frequency of "umbrella" tags may anticipate or possibly confirm and even numerically estimate the severity of rainy conditions. Rules of the type,

$$if\ (umbrella\ is\ HIGH-FREQUENCY)\ then\ (weather\ is\ VERY-RAINY)$$

may encode an association between the frequency of occurrence of the tag "umbrella" in messaging and the actual incidence of rain in a certain geographic location. Hence, an estimate of rain severity may be obtained, not from a rain gauge, but from word appearance and umbrella associations in the linguistic milieu of social media which may also include other modalities of language such as images, video, and of course numerical data coming either from sensors or from linguistic statements themselves.[1] It should be noted, however, that simple linguistic correlations, such as an association between the high frequency of umbrella tags and weather conditions, does not form a causal association in the sense that a naïve and nonsensical prediction may be made, which claims that umbrellas cause rain. The rules provide context and the essential public linguistic surroundings or social linguistic background for a language-game that models the phenomena of interest on the linguistic manifestations of social interaction activities.

A fuzzy *if/then* rule describes a relation amongst linguistic variables such as *weather* and *umbrellas* or *umbrellas* and *rain* above. Such rules are also called *fuzzy conditional statements* or simply *fuzzy implications*, in the sense that the truthfulness of veracity of the *LHS* implies something about the veracity of the *RHS*.[2]

Formally, given two different universes of discourse X and Y and two fuzzy variables x and y on X and Y, respectively, the assignment

---

[1]In the area of Big Data, in addition to the large volume of data, diverse sources of data are amalgamated including sources from gauges, sensors, and social media. Hence variety, along with volume, veracity, and velocity, form some characterizing features of Big Data.

[2]The use of the term *implication* is a bit of a misnomer. In mathematical logic, implication is the outcome of inferencing; a rule alone does not imply anything. But we will leave this for the purists and stay with the use of the term "implication" as a description of an association in a fuzzy rule.

of fuzzy value A on x and fuzzy value B on y leads to the elementary propositions:

$$p = x \text{ } is \text{ } A$$

and,

$$q = y \text{ } is \text{ } B$$

It should be noted that in propositions such as the ones above, the value A is assigned to the variable x. Hence, we refrain from denoting this as $x = A$, as it is sometimes done, because strictly speaking we are talking about an assignment and not equality.

If these propositions are associated through an *if/then* rule named R, we can write the rule R as

$$R\text{: } if \text{ } p \text{ } then \text{ } q$$

and, most commonly,

$$R: if \text{ } (x \text{ } is \text{ } A) \text{ } then \text{ } (y \text{ } is \text{ } B) \tag{3.1}$$

Rule (3.1) is the canonical form of fuzzy *if/then* rule, but alternative formulations are found in the literature such as

$$R\text{:} \quad A \to B$$

The expression $A \to B$ denotes an *if/then* rule with A and B as fuzzy sets indicating fuzzy values of x and y variables (not shown in these notation) found on the *LHS* and *RHS*. The relation that the rule captures is founded upon the Cartesian product $X \times Y$, where X and Y are two one-dimensional universes of discourse the elements x and y of which are arranged in ordered pairs.

The term *Cartesian product* refers to a higher dimensional universe of discourse, which is comprised of two or more universes of discourse endowed with structure, such as having elements from one universe of discourse associated with elements of another universe of discourse where the order of first and second is clearly designated as in the element of a pair (x, y) where x comes first and y comes second.

Cartesian products over two universes of discourse host *binary relations*, while three-dimensional universes of discourse serve as the foundation of *tertiary relations* and in general, a Cartesian product of n dimensions serves as the foundation of *n-ary relations*. Such relations have ordered structures called n-tuples, which are the generalization of pairs or triplets in binary and tertiary relations, respectively. When the elements of a relation belong to the relation fully, and those excluded do not belong at all, we have a *crisp relation*.

All relations are first and foremost defined linguistically before they can be represented mathematically in more abstract terms such as graphs, tables, and matrices. Their linguistic definition is based on some property or quality satisfied by pairs, triplets, or n-tuples.

From a linguistic definition, we can always extract a linguistic description in the form of *if/then* rules.

Consider for example the relation

$$R: \textit{\textbf{is\_a\_component\_of}}$$

The relation R may be used to describe the pair of elements (*battery, smart-phone*). Here it is important to observe the importance of order. If the linguistic description of

$$R: \text{ ``}\textit{battery } \textit{\textbf{is\_a\_component\_of}} \textit{ smart} - \textit{phone}\text{''}$$

is true, changing the order of the elements to

$$R^*: \text{ ``}\textit{smart} - \textit{phone } \textit{\textbf{is\_a\_component\_of}} \textit{ battery}\text{''}$$

gives a nonsensical statement. Comparing rule R to rule R^* we see that order matters, changing the order may completely change the meaning of the relation.

A relation such as "***is_a_component_of***" may "translated" to the canonical form of an *if/then* rule. In this format, we can rewrite R as:

$$R\!: \textit{if object is BATTERY then component\_of is SMART} - \textit{PHONE}$$

Where, *object* and *component_of* are fuzzy variables and *BATTERY* and *SMART – PHONE* are their corresponding values. Any ambiguity as to what degree an object is known to be a *battery* or a *smart-phone* or the degree of truth in the association results is a different kind of relation, no longer crisp, but fuzzy.

Relations are *mappings*. The archetypal mapping, which all scientists and engineers are familiar with being that of a function, are a special type of relation called *many-to-one* or *one-to-one*. A *many-to-one relation* is where many elements of the range (say elements $y_1, y_2, \ldots,$ on the y-axis or the Y universe of discourse) are associated with one (and only one) element x_0 of a domain that is the x-axis of a graph or the X universe of discourse, but not *vice versa*.

More general relations, however, are what is called *many-to-many* mappings. Many x's can be associated with a single y and *vice versa*. Many y's can also be associated with a single x. To be able to compute with such general relations is extremely important. It is really at the heart of turning to language as a form of computation.

Example 3.1 Different Representations of Relations Let us consider a *relation*, $R_{component\_of}$, on the Cartesian product $X \times Y$ formed between the universe of discourse X and the universe of discourse Y where a linguistic variable $x \in X$ may take the values:

$$X = \{BATTERY, CAMERA, GPS, CHIPS, WiFi\_ANTENNA\}$$

The linguistic variable $y \in Y$ may take on values

$$Y = \{CAR, SMART\_PHONE, HOME, LAB\}$$

The relation "x **is a component of** y". $R_{component\_of}$ is a *binary* relation because it involves two elements, x and y, drawn from the Cartesian product $X \times Y$. Furthermore, it is a *crisp* relation since the objects involved may be in a crisp "Yes" or "No" association. It is easy to list all the pairs of the relation and to see that the relation itself is a set, namely the crisp set of all the pairs

$$R_{component\_of} = \{(BATTERY, CAR), (BATTERY, SMART\_PHONE), (BATTERY, HOME), (CAMERA, SMART\_PHONE), (CAMERA, HOME), (GPS, SMART\_PHONE), (GPS, LAB), (CHIPS, SMART\_PHONE), (WiFi\_ANTENNA, SMART\_PHONE), (WiFi\_ANTENNA, LAB)\} \quad \text{(E3.1A)}$$

The relation $R_{component\_of}$ can also be represented through a graph as shown in Fig. 3.1. The individual elements are represented by oval nodes, called the *vertices* of the graph. If $R_{component\_of}$ is true for two elements we connect them by an arrow, the direction of the arrow indicating the order of the elements in the relation. For example, given that "3" divides "6" there is an arrow going from "3" to "6" and since "6" does not divide "3" there is no arrow going from "6" to "3." Reflecting the fact that the order of elements, or, the directions of the arrows, is important we call this a *directed graph*.

The binary relation $R_{component\_of}$ may also be represented by a table and a matrix. Table 3.1 shows the tabular representation of $R_{component\_of}$. When a table entry is "1" it indicates that x (row entry) is a component of the item in the corresponding y (column entry) as for example in the fourth row and second column we simply have that the chips are components of smart phones. A "0" indicates the absence of such a relation.

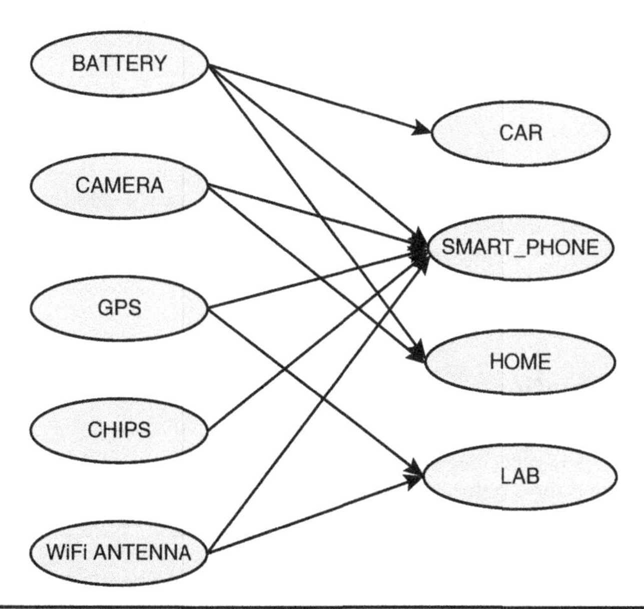

Figure 3.1 The directed graph of the relation $R_{component\_of}$ defined on the Cartesian product $X \times Y$.

$R_{component\_of}$:	*y*	*CAR*	*SMART_ PHONE*	*HOME*	*LAB*
	x				
	BATTERY	1	1	1	1
	CAMERA	0	1	1	1
	GPS	0	1	0	0
	CHIPS	0	1	0	1
	WiFi ANTENNA	0	1	0	0

TABLE 3.1 A tabular representation of the relation $R_{component\_of}$ of Example 3.1

$R_{component\_of}$ can also be represented by a matrix obtained from Table 3.1 by removing the column of x's on the side and the row of y's from the top, i.e.,

$$R_{component\_of} = \begin{bmatrix} 1 & 1 & 1 & 1 \\ 0 & 1 & 1 & 1 \\ 0 & 1 & 0 & 0 \\ 0 & 1 & 0 & 0 \\ 0 & 1 & 0 & 0 \end{bmatrix} \qquad \text{(E3.1B)}$$

Thus, we have seen five different ways for representing $R_{component\_of}$:

1. *Linguistically*, through the statement "$R_{component\_of}$"
2. By listing the *set* of all ordered pairs as in Eq. (E3.1A)
3. As a *directed graph* (Fig. 3.1)
4. As a *table* (Table 3.1), and finally
5. As a *matrix*, Eq. (E3.1B)

It should be noted that the last two ways are generally convenient only for *binary* relations. For *tertiary* relations, for example, we would need a three-dimensional table or matrix, (for *n-ary* relations *n*-dimensional tables and matrices) and therefore tables and matrices may be conveniently used only with *binary* relations.◊

3.2 Fuzzy Relations

In fuzzy relations, the elements of a relation, that is, *pairs* or more generally *n*-tuples, have degrees of membership in a certain relation, that is, they belong to *a degree*. Just as the question of whether some element belongs to a set may be considered a matter of degree, whether some elements are associated or not may also be a matter of degree (Zadeh, 1971; (Dubois, 1980). Thus, fuzzy relations are high-dimensional fuzzy sets established on Cartesian products.

A Cartesian product as such is simply a higher dimensional universe of discourse.

A fuzzy relation R defined on $X \times Y$ is a fuzzy set for which we can list all pairs explicitly; every pair needs to have its membership in the relation stated, that is, be a singleton $\mu_R(x, y)/(x, y)$, and the relation is the union of all the singletons of $X \times Y$. For a discrete relation, that is, a discrete Cartesian product, we would have

$$R = \sum_{(x_i, y_j) \in X \times Y} \mu_R(x_i, y_j)/(x_i, y_j) \tag{3.2}$$

while for a continuous Cartesian product, we have

$$R = \int_{X \times Y} \mu_R(x_i, y_j)/(x_i, y_j) \tag{3.3}$$

The same notation is used for any *n-ary fuzzy relation.*

Like with crisp relations, we use five distinct, inequivalent, yet equally valid ways of representing fuzzy relations.

1. *Linguistically*, through fuzzy *if/then* rules
2. As a fuzzy set, that is, listing the *set* of all singletons
3. As a *directed graph*
4. As a *table*
5. As a *matrix*

Of special interest to us are the matrix representations of relations. This is because linguistic descriptions, that is, collections of *if/then* rules, are a façade. Underneath them there are matrix representations and operations amenable to computing. Computers like to crunch numbers and humans like to use words. Fuzzy logic bridges the two (especially language and AI) by providing a mathematical substrate upon which to establish and use language-games (that is, clusters of *if/then* rules).

Let us now look at some matrices of special interest in undergirding fuzzy relations. The *membership matrix* of an $n \times m$ binary fuzzy relation has the general form:

$$R = \begin{bmatrix} \mu_R(x_1, y_1) & \mu_R(x_1, y_2) & \cdots & \mu_R(x_1, y_n) \\ \mu_R(x_2, y_1) & \mu_R(x_2, y_2) & \cdots & \mu_R(x_2, y_n) \\ & \vdots & & \\ \mu_R(x_m, y_1) & \mu_R(x_m, y_2) & \cdots & \mu_R(x_m, y_n) \end{bmatrix} \tag{3.4}$$

The *Identity Fuzzy Relation*, R_I, represents a special type of membership relation which has "1" in all diagonal elements and "0" in all off-diagonal elements, i.e.,

$$R_I = \begin{bmatrix} 1 & 0 & & 0 \\ 0 & 1 & & 0 \\ & & \ddots & \\ 0 & 0 & & 1 \end{bmatrix} \tag{3.5}$$

Another special relation is the *Universe Relation*, R_E, namely a relation with "1" everywhere in its membership matrix. It is a representation of the universe of discourse or a Cartesian product, i.e.,

$$R_E = \begin{bmatrix} 1 & 1 & \cdots & 1 \\ 1 & 1 & & 1 \\ \vdots & & \ddots & \\ 1 & 1 & & 1 \end{bmatrix} \tag{3.6}$$

The *Null Relation*, R_0, has a membership matrix with "0" everywhere, i.e.,

$$R_0 = \begin{bmatrix} 0 & 0 & \cdots & 0 \\ 0 & 0 & & 0 \\ \vdots & & \ddots & \\ 0 & 0 & & 0 \end{bmatrix} \tag{3.7}$$

The transpose of a membership matrix gives the membership matrix of the *inverse relation* of R denoted by R^{-1}, and defined by

$$\mu_{R^{-1}}(y, x) \equiv \mu_R(x, y) \tag{3.8}$$

Thus, the *inverse* of the relation represented by the matrix of Eq. (3.8) has the membership matrix

$$R^{-1} = \begin{bmatrix} \mu_R(x_1, y_1) & \mu_R(x_2, y_1) & \cdots & \mu_R(x_m, y_1) \\ \mu_R(x_1, y_2) & \mu_R(x_2, y_2) & \cdots & \mu_R(x_m, y_2) \\ & \vdots & & \\ \mu_R(x_1, y_n) & \mu_R(x_2, y_n) & \cdots & \mu_R(x_m, y_n) \end{bmatrix} \tag{3.9}$$

which is the transpose of the matrix found by interchanging the rows of R to produce the columns of R^{-1} and the columns of R have become the rows of R^{-1} (Terano, 1992). The inverse of an inverse relation is the original relation just as the inverse of the inverse of a matrix is the original matrix, i.e.,

$$(R^{-1})^{-1} = R \tag{3.10}$$

So far, we defined fuzzy relations on crisp Cartesian products. However, fuzzy relations can also be defined on fuzzy Cartesian products (Tsoukalas, 1997).

3.3 Operators for Implications and Connectives

The membership function that defines the fuzzy relation of an *if/then* rule can be obtained from the individual membership functions of the left- and right-hand side of a rule. For instance, if the *if/then* rule is

$$R:\quad if\ x\ is\ A\ then\ y\ is\ B \tag{3.11}$$

Then, the membership function of R, that is, $\mu_R(x,y)$, can be obtained from the membership functions of the *LHS*, $\mu_A(x)$, and the *RHS*, $\mu_B(y)$, and furthermore, this can happen in several different ways (Tsoukalas, 1997). For convenience, we introduce an *implication operator*, φ_α, that is, a mathematical mechanism that takes as input the relevant membership functions of the *LHS* and the *RHS* and outputs the membership function $\mu_R(x,y)$ of the relation. Hence if we assume that

$$\mu_R(x,y) = \varphi_\alpha\,[\mu_A(x),\,\mu_B(y)],\quad \forall\ x \in X \text{ and } y \in Y \tag{3.12}$$

After obtaining the membership function of a rule, the fuzzy relation that describes the rule is:

$$R = \int_{X\times Y} \mu_R(x_i,\,y_j)\,/\,(x_i,\,y_j),\quad \forall\ (x,y) \in X \times Y \tag{3.13}$$

The implication operator φ_α is selected from a list of options, as shown in Table 3.2, where we list the 10 most commonly used operators. In engineering applications such as control, the first two operators, that is, Mamdani, ϕ_c, and Larsen, ϕ_p, are used most often. One interesting feature is that the choice of the fuzzy implication operator φ_α determines the interpretation of the *ELSE* connective, also obtained through a *connective operator* φ_β, and applicable in situations where we have clusters or rules working together. The rules in a cluster represent a set of linguistic descriptions to achieve a task, such as a control objective. A cluster of rules, R^C, also referred to as a *rule block*, or, *fuzzy algorithm*, constitutes the core of a particular language-game.

In a cluster of rules, if an *LHS* variable x takes on values $A^1,\ A^2,\ldots,A^n$ and the *RHS* variable y takes on values $B^1,\ B^2,\ldots,B^n$ we can assemble several *if/then* rules to work together as a fuzzy algorithm. The rules of the algorithm are connected with each other through a connective *ELSE*, which has a *connective operator*, φ_β, that signifies that the rules work together toward a specific task or purpose. Thus, the choice of implication operator, φ_α, determines the model of an *if/then* rule and gives us the membership function of the rule in terms of the membership functions of its constituent parts. On the other hand, the choice of implication operator φ_α also determines the operator ϕ_β for the connective *ELSE* as shown in the last column of Table 3.2.

Name	φ_α, Implication Operator	ϕ_β, ELSE
ϕ_c, *Mamdani*	$\mu_A(x) \wedge \mu_B(y)$	*OR* ($\vee$)
ϕ_p, *Larsen*	$\mu_A(x) \cdot \mu_B(y)$	*OR* ($\vee$)
ϕ_m, *Zadeh*	$(\mu_A(x) \wedge \mu_B(y)) \vee (1 - \mu_A(x))$	*AND* ($\wedge$)
ϕ_a, *Arithmetic*	$1 \wedge (1 - \mu_A(x) + \mu_B(y))$	*AND* ($\wedge$)
ϕ_b, *Boolean*	$(1 - \mu_A(x)) \vee \mu_B(y)$	*AND* ($\wedge$)
ϕ_{bp}, *Bounded Product*	$0 \vee (\mu_A(x) + \mu_B(y) - 1)$	*OR* ($\vee$)
ϕ_{dp}, *Drastic Product*	$\mu_A(x),\ iff\ \ \mu_B(y) = 1$ $\mu_B(y),\ iff\ \ \mu_A(x) = 1$ $0,\ iff\ \ \mu_A(x) < 1,\ \mu_B(y) < 1$	*OR* ($\vee$)
ϕ_s, *Standard Sequence*	$1,\ iff\ \ \mu_A(x) \leq \mu_B(y)$ $0,\ iff\ \ \mu_A(x) \geq \mu_B(y)$	*AND* ($\wedge$)
ϕ_Δ, *Gougen*	$1,\ iff\ \ \mu_A(x) \leq \mu_B(y)$ $\frac{\mu_B(y)}{\mu_A(x)},\ iff\ \ \mu_A(x) > \mu_B(y)$	*AND* ($\wedge$)
ϕ_g, *Gödelian*	$1,\ iff\ \ \mu_A(x) \leq \mu_B(y)$ $\mu_B(y),\ iff\ \ \mu_A(x) > \mu_B(y)$	*AND* ($\wedge$)

TABLE 3.2 Ten Most Common Implication Operators φ_α and *ELSE* Operator ϕ_β, Used in Modeling Fuzzy *if/then* Rules

Consider, for instance, the cluster of n rules R^N given below:

$$R^N:\ \ if\ x\ is\ A^1\ then\ y\ is\ B^1, ELSE$$
$$if\ \ x\ is\ A^2\ \ then\ y\ is\ B^2, ELSE$$
$$\ldots\ldots\ldots\ldots$$
$$if\ y\ is\ A^n\ then\ y\ is\ B^n \tag{3.14}$$

The operator for the connective *ELSE*, denoted by φ_β, is limited by the choice of implication operator φ_α used in the definition (or modeling) of a particular *if/then* rule. Therefore, for the fuzzy algorithm of Eq. (3.4), we have

$$\mu_{R^N}(x,y) = \varphi_\beta[\mu_{R^1}(x,y),\ \ \mu_{R^2}(x,y), \ldots, \mu_{R^n}(x,y)]$$

or,

$$\mu_{R^N}(x,y) = \varphi_\beta\ [\varphi_\alpha\ (\mu_{A^1}(x)\ ,\ \mu_{B^1}(y)),$$
$$\varphi_\alpha\ (\mu_{A^2}(x),\ \mu_{B^2}(y)), \ldots, \varphi_\alpha\ (\mu_{A^n}(x),\ \mu_{B^n}(y))] \tag{3.15}$$

The equations above were developed for simple *if/then* rules each having a *LHS* variable x taking on a value A and a *RHS* variable y taking on a value B.

In general, any language-game can be reduced to a collection of simple *if/then* rules. Consider, for example, a more complex rule structure where the *LHS* contains more than one proposition in a nested structure of conditionals which may reflect a language game of discovering potable water in the desert island though experiment of Chap. 1 and may go like this "if we find a stream of running water and observe animals drinking it then this water could be safe to drink." In such a case, the condition part (*LHS*) may be considered as a series of embedded dependent fuzzy relations

$$\begin{aligned} &\text{if } x_1 \text{ is } A^1 \text{ then} \\ &\quad (\text{if } x_2 \text{ is } A^2 \text{ then} (\text{if } x_3 \text{ is } A^3 \text{ then}.... (\text{if } x_m \text{ is } A^m \text{ then } y \text{ is } B))...) \end{aligned} \tag{3.16}$$

or as a relation where the variables of antecedents (*LHS*) are connected with the connective *AND* and take the form

$$\begin{aligned} &\text{if } (x_1 \text{ is } A^1 \text{ AND } x_2 \text{ is } A^2 \\ &\quad \text{AND } x_3 \text{ is } A^3 \ldots \text{AND } x_m \text{ is } A^m) \text{ then } y \text{ is } B \end{aligned} \tag{3.17}$$

Hence, in the former case we may have as an analytical (mathematical) substrate for the linguistic description of Eq. (3.18)

$$\begin{aligned} &\mu_R(x_1,x_2,\ldots,x_m,y) \\ &= \varphi_\alpha(\mu_{A^1}(x_1), \quad \varphi_\alpha\,(\mu_{A^2}(x_2), \\ &\varphi_\alpha(\mu_{A^3}(x_3), \ldots.\ \varphi_\alpha(\mu_{A^m}(x_m),\ \mu_B(y))))), \\ &\forall\ x_1,x_2,\ldots,x_m\ \in X_1,\ X_2,\ldots,X_m,\ \text{ and } y \in Y \end{aligned} \tag{3.18}$$

or,

$$\begin{aligned} &\mu_R(x_1, x_2, \ldots, x_m, y) \\ &= \varphi_\alpha\{\mu_{A^1}(x_1) \wedge \mu_{A^2}(x_2) \wedge \mu_{A^3}(x_3)\ldots\mu_{A^m}(x_m),\ \mu_B(y)\} \\ &= \varphi_\alpha \left\{ \bigwedge_{k=1}^{m} \mu_{A^k}(x_k),\ \mu_B(y) \right\} \\ &\forall\, x_k \in X_k \text{ and } y \in Y \end{aligned} \tag{3.19}$$

In general, the above equations are realized mathematically in different ways depending on the operator φ_α but for some operators φ_α they are equivalent.

Let us next turn our attention to how different implication operators can be used in fuzzy algorithms. For example, if *Mamdani implication* is used, we have

$$R: \quad \text{if } x \text{ is } A \quad \text{then } y \text{ is } B \tag{3.20}$$

with membership function for the underlying fuzzy relation given by

$$\mu_R(x,y) = \varphi_C[\mu_A(x),\ \mu_A(x)] = \mu_A(x) \wedge\ \mu_A(x)$$

Alternatively we can write an *if/then* rule (3.22) under Mamdani implication using the shorthand

$$A \to B \equiv A \wedge B$$

This shorthand indicates that the rule has membership function obtained from the membership functions of A and B by taking the minimum of the two. It is shorthand notation, which is convenient when we have chains of rules, such as

$$\begin{aligned} A_1 \to A_2 \to B &= A_1 \wedge (A_2 \wedge B) \\ &= (A_1 \wedge A_2) \wedge B \\ &= (A_1 \wedge A_2) \to B \end{aligned} \tag{3.21}$$

If we use the *Boolean implication operator* (see Table 3.1), we can alternatively write for (3.12)

$$A \to B \equiv \bar{A} \vee B,$$

$$\begin{aligned} A_1 \to A_2 \to B &= A_1 \to (\bar{A}_2 \vee B) \\ &= \bar{A}_1 \vee (\bar{A}_2 \vee B) \\ &= (\bar{A}_1 \vee \bar{A}_2) \vee B \\ &= \overline{(A_1 \wedge A_2)} \vee B \\ &= (A_1 \wedge A_2) \to B \end{aligned} \tag{3.22}$$

If the Arithmetic implication operator, φ_α, is used, the logical equivalences obtained in Eqs. (3.22) are generally not true. Hence we have to choose *if/then* structures of the form of Eqs. (3.16) or (3.17), a choice made based on the semantics of the application. For the convenience of implementation, Eq. (3.17) is usually easier to implement. Both relations can be easily realizable when we use the Larsen product implication operator which gives us

$$\mu_R(x_1, x_2, \ldots, x_m, y) = \mu_{A^1}(x_1) \wedge \mu_{A^2}(x_2) \wedge \mu_{A^3}(x_3) \ldots \mu_{A^m}(x_m), \mu_B(y)$$

$$= \prod_{k=1}^{m} \mu_{A^k}(x_k) \cdot \mu_B(y)$$

$$\mu_R(x_1, x_2, \ldots, x_m, y) = \bigwedge_k^m \mu_{A^k}(x_k) \cdot \mu_B(y) \tag{3.23}$$

Two fuzzy *if/then* rules in a fuzzy algorithm may be merged into one, in order to simplify things. The merging takes place and results in one fuzzy *if/then* rule if the initial two rules differ in no more than one *LHS* proposition. For example, if we have two rules with the same *RHS*, but their *LHS* is different only in one proposition, that is, the first has x *is* A_1 as antecedent, while the second has only x *is* A_2 as antecedent, as shown below,

$$\begin{aligned} &if\ x\ is\ A_1\ then\ (if\ y\ is\ B\ then\ z\ is\ C) \\ &if\ x\ is\ A_2\ then\ (if\ y\ is\ B\ then\ z\ is\ C) \end{aligned} \qquad (3.24)$$

Then the two rules can be merged into one rule which takes the form,

$$if\ (x\ is\ A_1\ OR\ x\ is\ A_2)\ then\ (if\ y\ is\ B\ then\ z\ is\ C) \qquad (3.25)$$

Rule (3.25) is semantically equivalent to rules (3.24), although the two formulations are different in appearance (syntactically different).

A similar thing happens in fuzzy *if/then* rules below:

$$\begin{aligned} &if\ x\ is\ A\ AND\ y\ is\ B_1\ then\ z\ is\ C \\ &if\ x\ is\ A\ AND\ y\ is\ B_2\ then\ z\ is\ C \end{aligned} \qquad (3.26)$$

These two rules can be merged in a single rule,

$$if\ x\ is\ A\ AND\ ((y\ is\ B_1)\ OR\ (y\ is\ B_2))\ then\ z\ is\ C \qquad (3.27)$$

The compound (composite) terms shown in the *LHS* of (3.27) using *OR* and *AND* can be calculated as we have shown in Chap. 2.

Now, let as turn our attention in more detail to some of the definitions of the operator, φ_α, through which we obtain the fuzzy relation that undergirds the fuzzy *if/then*. When the rule R has the form of (3.11), that is,

$$R\colon if\ x\ is\ A\ then\ y\ is\ B$$

the Zadeh implication operator φ_m gives us the membership function of the rule as (see Table 3.2)

$$\mu_R(x,y) = \varphi_m[\mu_A(x), \mu_B(y)] = (\mu_A(x) \wedge \mu_B(y)) \vee (1-\mu_A(x)) \qquad (3.28)$$

The simpler Mamdani φ_c, on the other hand, give us

$$\mu_R(x,y) = \varphi_c[\mu_A(x), \mu_B(y)] = \mu_A(x) \wedge \mu_B(y) \qquad (3.29)$$

Combining n fuzzy *if/then* rule relations can be achieved through the connective *OR*, that is,

$$R^N = \bigvee_{j=1}^{n} R^j$$

and,

$$\mu_{R^N}(x,y) = \bigvee_{j=1}^{n} (\mu_{A^j}(x) \wedge \mu_{B^j}(y)) \qquad (3.30)$$

As another example, consider the cases of rules where we use Boolean implication φ_b. In this case, the membership function of the relation is

$$\mu_R(x,y) = \varphi_b[\mu_A(x), \mu_B(y)] = (1 - \mu_A(x)) \vee \mu_B(y)$$

The combination of n fuzzy *if/then* rules takes place with the connective *AND* (as seen in Table 3.2), that is,

$$R^N = \bigwedge_{j=1}^{n} R^j$$

and

$$\mu_{R^N}(x, y) = \bigwedge_{j=1}^{n} [(1 - \mu_{A^j}(x)) \vee \mu_{B^j}(y)] \tag{3.31}$$

The arithmetic implication operator φ_a which is based in the multi-valued logic of Lukasiewicz, was given to us in Zadeh (1975), and results in

$$\mu_R(x,y) = \varphi_a[\mu_A(x), \mu_B(y)] = 1 \wedge (1 - \mu_A(x)) + \mu_B(y)$$

where, the + in the above equation indicates addition (as in normal arithmetic). The combination of N *if/then* rules in a fuzzy algorithm used the connective *AND*($\wedge$), that is,

$$R^N = \bigwedge_{j=1}^{n} R^j$$

$$\mu_{R^N}(x,y) = \bigwedge_{j=1}^{n} [1 \wedge (1 - \mu_{A^j}(x) + \mu_{B^j}(y)] \tag{3.32}$$

The Larsen implication operator, φ_p, uses ordinary arithmetic product and it is found in numerous successful control applications. The membership function of the *if/then* rule is obtained as:

$$\mu_R(x,y) = \varphi_p[\mu_A(x), \mu_B(y)] = \mu_A(x) \cdot \mu_B(y)$$

The combination of N *if/then* rules is performed using *OR*($\vee$), that is,

$$R^N = \bigvee_{j=1}^{N} R^j$$

and,

$$\mu_{R^N}(x,y) = \bigvee_{j=1}^{n} (\mu_{A^j}(x) \cdot \mu_{B^j}(y)) \tag{3.33}$$

From the menu of implication operators given in Table 3.1, including the concomitant operators for the connectives *ELSE* which cement different rules in rule clusters and the analysis above, we see that the simplest implication operators computationally are Mamdani and

Larsen. Fortunately, in most applications these are sufficient for modeling purposes, especially because in the operation of composition they give us very simple but effective means of evaluating *if/then* rules.

3.4 Composition and the Compositional Rule of Inference

Composition is a powerful tool for evaluating relations which are not limited to the usual many-to-one or one-to-one variety of algebra and calculus. In the case of fuzzy relations, which are inherently *many-to-many* mappings, composition helps us to evaluate them easily and efficiently. Since fuzzy relation underline the logical façade of fuzzy *if/then* rules, composition is the mathematical operation underlying logical inferencing procedures such as *Generalized Modus Ponens* (*GMP*) and *Generalized Modus Tollens (GMT)* which are fuzzy extensions of the two best known syllogisms of Aristotelian logic (*modus ponens* and *modus tollens*). GMP is used extensively for evaluating individual if/then rules as well as clusters of rules.

Suppose that we have two *if/then* rules

$$R^1: \text{ } if \ A \ then \ B$$
$$R^2: \text{ } if \ B \ then \ C$$

They can be composed into one conditional relation, that is,

$$R^{12}: \text{ } if \ A \ then \ C$$

The composition

$$R^{12} = R^1 \circ R^2$$

can be performed, either through *max-min*, that is,

$$\mu_{R^{12}}(x,y) = \bigvee_y (\mu_{R^1}(x,\ y) \wedge \mu_{R^2}(y,\ z)) \tag{3.34}$$

or, through *max-product*, that is,

$$\mu_{R^{12}}(x,y) = \bigvee_y (\ \mu_{R^1}(x,\ y) \cdot \mu_{R^2}(y,\ z)) \tag{3.35}$$

When we use discrete fuzzy sets, the operations above turn out to be equivalent to the matrix product of matrices, provided that multiplication is substituted with min and addition with max operators. It makes eminent sense to propose the use of the two definitions regarding the definition of implication $\mu_R(x,y)$.

As a general rule, with implication operators that involve max and min, we use the compositional rule of inference max-min. With implication operators that involve arithmetic operators (and especially the implication Max-Product), we can use the compositional rule of inference max-product.

The problem that we will examine now is finding the *consequent* for a given *antecedent* when we have the relation amongst two fuzzy variables. If,

$$A = \{\mu_A(x)/x\}, \quad x \in X$$
$$B = \{\mu_B(y)/y\}, \quad y \in Y$$

And, after we apply an implication operator, we obtain the two-dimensional implication relation

$$R = \{\mu_R(x, y)/(x,y)\}, \quad (x,y) \in X \times Y$$

Then, if the antecedent A' (cause) is

$$A' = \{\mu_{A'}(x)/x\}, \quad x \in X$$

The consequent B' is inferred using the compositional rule of inference (Zadeh, 1973), that is,

$$B' = A' \circ R$$

or,

$$B' = A' \circ R = \left\{ \bigvee_x (\mu_{A'}(x) \wedge \mu_{R^2}(y, z))/y \right\} \tag{3.36}$$

Equation (3.36) is the usual way we evaluate through composition fuzzy *if/then* rules, which are activated by an input A' and we wish to figure out the output B'. We recall that this process is logically called *GMP* after the famous Aristotelian syllogism (Modus Ponens = the way of the bridge) generalized for fuzzy operations.

In *GMP* using composition, a significant difference is made by the choice of implication operators used to model the fuzzy *if/then* rule. If *Mamdani* φ_C (Table 3.2) is used the result entirely depends on the *Degree of Fulfillment* (*DOF*) of the rule for a given input A'. The *DOF* is a measure of the degree to which the input A' matches the *LHS* of the rule. Under Mamdani, the *RHS* of the rule gets clipped by the DOF to produce a resultant membership function for B′. Under Larsen φ_p (Table 3.2), the *DOF* scales (multiplies) the *RHS*. In the examples that follow we look at these two different effects that implication operators produce when rules are evaluated in *GMP* using composition.

Example 3.2 Using Max-Min Composition for Rules under Mamdani Implication In this example, we use max min composition, see Eq. (E3.36), to evaluate a rule when it is presented with an input value at some given time. The fuzzy values A and B on the *LHS* and *RHS* sides of an *if/then* rule

$$\textit{if } x \textit{ is } A \textit{ then } y \textit{ is } B \tag{E3.2A}$$

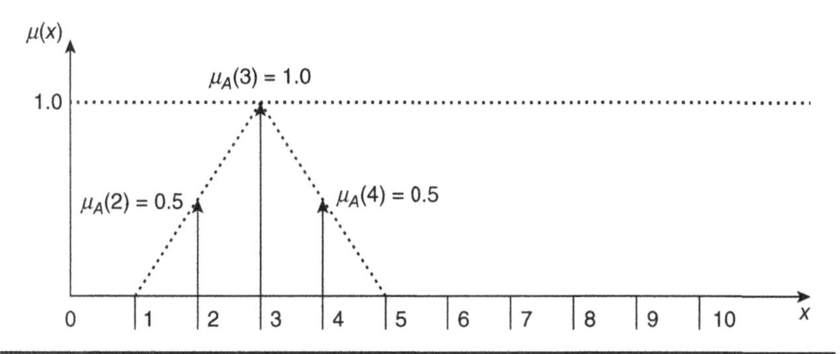

FIGURE 3.2 The membership function of *A* in the rule (E3.2A).

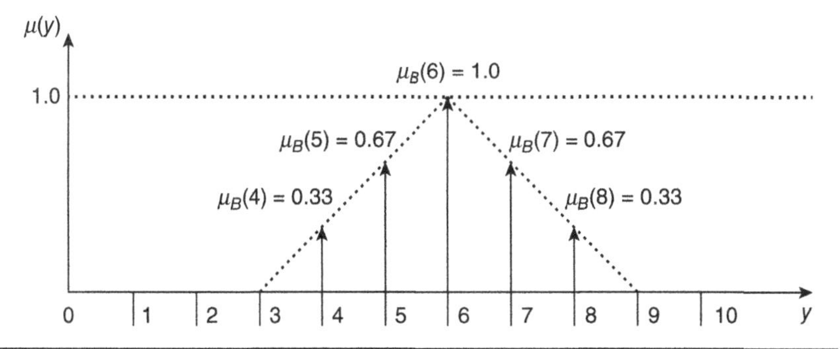

FIGURE 3.3 The membership function of *B* in the rule (E3.2A).

are given below and are graphically depicted in Zadeh diagrams in Figs. 3.2 and 3.3.

$$A = \sum_{i=0}^{10} \mu_A(x_i)/x_i = 0.5/2 + 1.0/3 + 0.5/4$$

$$B = \sum_{i=0}^{10} \mu_B(y_i)/y_i = 0.33/4 + 0.67/5 + 1.0/6 + 0.67/7 + 0.33/8$$

If the rule in (E3.2A) is modeled under Mamdani implication operator φ_C (Table 3.1), the fuzzy relation for the rule is calculated to be

$$\begin{aligned} R(x_i, y_i) = & \sum_{(x_i, y_i)} \mu\,(x_i, y_j)/(x_i, y_j) \\ & = 0.33/(2,4) + 0.5/(2,5) + 0.5/(2,6) + 0.5/(2,7) + 0.33/(2,8) \\ & \quad + 0.33/(3,4) + 0.67/(3,5) + 1.0/(3,6) + 0.67/(3,7) + 0.33/(3,8) \\ & \quad + 0.33/(4,4) + 0.5/(4,5) + 0.5/(4,6) + 0.5/(4,7) + 0.33/(4,8) \end{aligned}$$

Let us assume an input A' (shown in Fig. 3.4 and given in the following text) is presented to the rule

$$A' = \sum_{i=0}^{10} \mu_{A'}(x_i)/x_i = 1.0/3$$

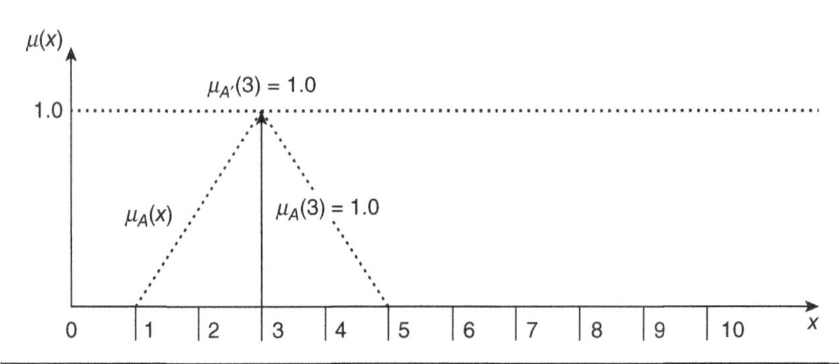

FIGURE 3.4 The input A' to the rule is the singleton shown.

We use *max-min composition* given in equation (3.36) to compute the output B' as follows:

$$B'(y_j) = A'(x_i)^\circ R(x_i, y_j)$$

$$= [0\ 1\ 0]^\circ \begin{bmatrix} 0.33 & 0.5 & 0.5 & 0.5 & 0.33 \\ 0.33 & 0.67 & 1.0 & 0.67 & 0.33 \\ 0.33 & 0.5 & 0.5 & 0.5 & 0.33 \end{bmatrix} \quad \text{(E3.2B)}$$

We recall, that max-min composition is like matrix multiplication, except we use the *max* (∨) operator in place of addition, and the *min* (∧) operator in place of multiplication. For example, to calculate the degree of membership of the number 4 to the set B' we "multiply" the row vector to the left with the first column vector in (E3.2B) above. This operation would give us

$$\mu_{B'}(4) = \bigvee_x [0 \wedge 0.33, 1 \wedge 0.33, 0 \wedge 0.33] = \bigvee_x [0, 0.33, 0] = 0.33$$

Hence, the membership of 4 to the set B' is 0.33. Similarly, we perform the operations with the other columns of the membership matrix in (3.38) to obtain the entire B':

$$B' = 0.33/4 + 0.67/5 + 1.0/6 + 0.67/7 + 0.33/8$$

The output set B' of the composition is shown in Fig. 3.5, and we note that it is the entire *RHS* value *B*. This is due to the fact that the degree of *fulfillment* of the

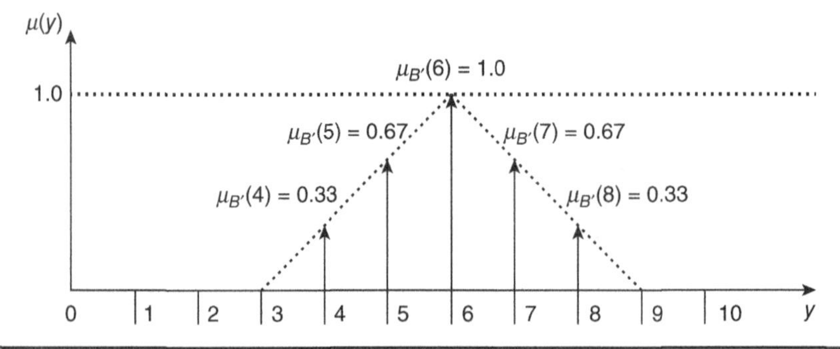

FIGURE 3.5 The output B' to the rule is the same as the *RHS* value of *B* since the *DOF* = 1.

rule in this example is $DOF = 1$, which means that the *RHS* fuzzy value of the rule will be "clipped" at a height of 1, that is, it will give the entire *RHS* value B as output B'.

Example 3.3 Using Max-Min Composition for Rules under Larsen Implication In this example we use max min composition see Eq. (3.36) to evaluate a rule modeled by Larsen implication (φ_p), as shown in Table 3.2. The fuzzy values A and B on the *LHS* and *RHS* sides of an *if/then* rule

$$\textit{if } x \textit{ is } A \textit{ then } y \textit{ is } B \tag{E3.3A}$$

are given below and are graphically depicted in Zadeh diagrams in Figs. 3.6 and 3.7.

$$A = \sum_{i=-5}^{i=5} \mu_A(x_i)/x_i = 0.33/(-2) + 0.67/(-1) + 1.0/0 + 0.75/1 + 0.5/2 + 0.25/3$$

$$B = \sum_{i=-5}^{i=5} \mu_B(y_i)/y_i = 0.50/(-3) + 1.0/(-2) + 0.67/(-1) + 0.33/0$$

If Larsen implication is used, the fuzzy relation modeling rule (E3.3A) is

$$R(x_i, y_i) = \sum_{(x_i, y_i)} \mu\,(x_i, y_j)/(x_i, y_j)$$
$$= 0.165/(-2,-3) + 0.33/(-2,-2) + 0.221/(-2,-1) + 0.109/(-2,0)$$

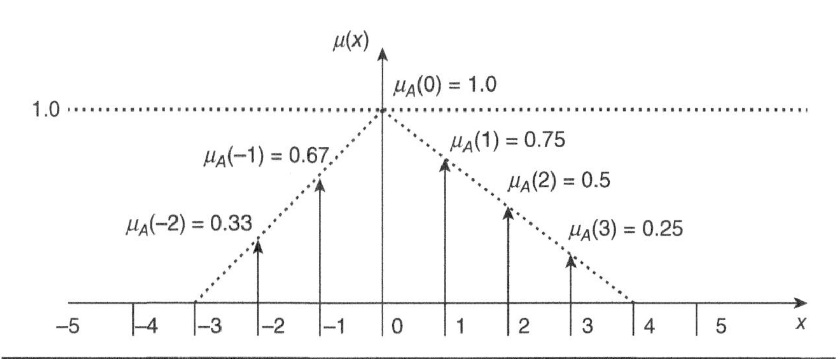

Figure 3.6 The *LHS* value *A* of the rule in Example 3.3.

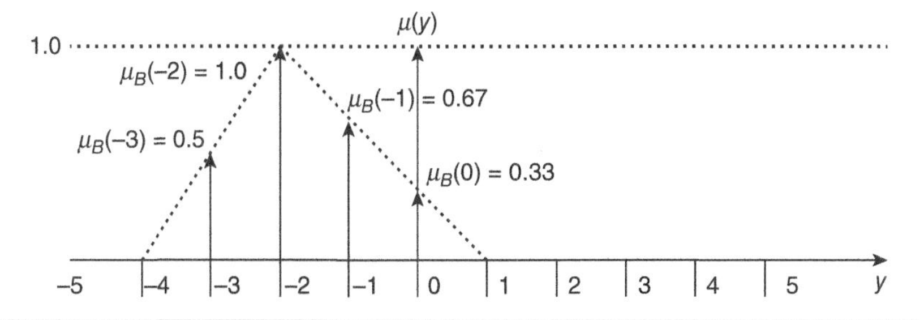

Figure 3.7 The *RHS* value *B* of the rule in Example 3.3.

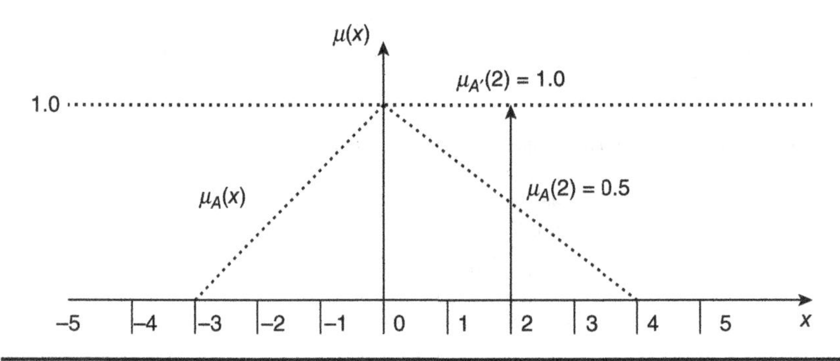

FIGURE 3.8 The input value A' given as input to the rule of Example 3.3.

$$
\begin{aligned}
&+0.335/(-1,-3)+0.67/(-1,-2)+0.301/(-1,-1)+0.221/(-1,0)\\
&+0.50/(0,-3)+1.00/(0,-2)+0.67/(0,-1)+0.33/(0,0)\\
&+0.375/(1,-3)+0.75/(1,-2)+0.502/(1,-1)+0.247/(1,0)\\
&+0.25/(2,-3)+0.5/(2,-2)+0.335/(2,-1)+0.165/(2,0)\\
&+0.125/(3,-3)+0.25/(3,-2)+0.167/(3,-1)+0.083/(3,0)
\end{aligned}
$$

Next, let us give the input A' to the rule (E3.3A) which is shown below and is graphically depicted in the Zadeh diagram of Fig. 3.8

$$A' = \sum_{i=-5}^{i=5} \mu_{A'}(x_i)/x_i = 1.0/2$$

To compute the output B' we perform max-min composition as described in Eq. (3.36), which results in the following:

$$
\begin{aligned}
B'(y_j) &= A'(x_i)^\circ R(x_i, y_j)\\
&= [0\ 0\ 0\ 0\ 1\ 0]^\circ \begin{bmatrix} 0.165 & 0.33 & 0.221 & 0.109\\ 0.335 & 0.67 & 0.301 & 0.221\\ 0.5 & 1.00 & 0.67 & 0.33\\ 0.375 & 0.75 & 0.502 & 0.247\\ 0.25 & 0.5 & 0.335 & 0.165\\ 0.125 & 0.25 & 0.167 & 0.083 \end{bmatrix}
\end{aligned} \tag{E3.3B}
$$

Max-min composition is like matrix multiplication, except we use the *max* (∨) operator in place of addition, and the *min* (∧) operator in place of multiplication. For instance, the degree of membership of the number –3 to the set B', that is, $\mu_{B'}(-3)$ would be calculated in the max-min composition of (E3.3B) as follows:

$$
\begin{aligned}
\mu_{B'}(-3) &= \bigvee_x [0 \wedge 0.165,\ 0 \wedge 0.335, 0 \wedge 0.5, 0 \wedge 0.375,\ 1 \wedge 0.25,\ 0 \wedge 0.125]\\
&= \bigvee_x [0,\ 0,\ 0,\ 0,\ 0.25,\ 0] = 0.25
\end{aligned}
$$

Hence, the entire B' output is

$$B' = 0.25/-3 + 0.5/-2 + 0.335/-1 + 0.165/0 \tag{E3.3C}$$

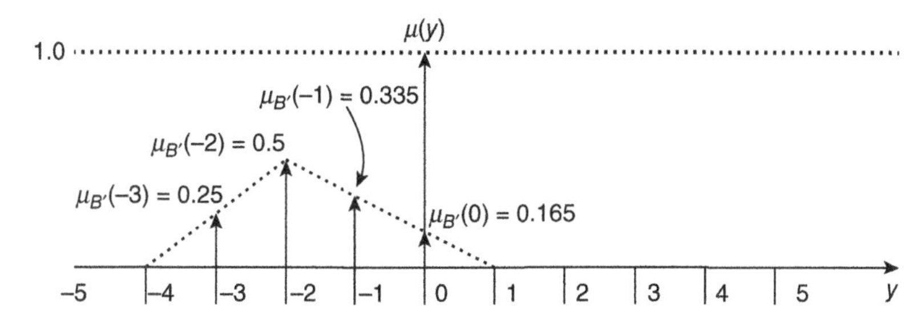

FIGURE 3.9 The output B' to the rule is scaling *RHS* value of B with the $DOF = 0.5$.

It should be noted by observing (E3.3C) and Fig. 3.9 that the output B' is a scaled version of B where the scaling is performed by multiplying B with the degree of fulfillment of the rule. In this case $DOF = 0.5$ resulting in a B' which is a B at half its original height.

References

Dubois, D., and Prade, H., *Fuzzy Sets and Systems: Theory and Applications,* Academic Press, Boston, 1980.

Terano, T., Asai, K., and Sugeno, M., *Fuzzy Systems Theory and Its Applications,* Academic Press, Boston, 1992.

Tsoukalas, L. H., and Uhrig, R. E., *Fuzzy and Neural Approaches in Engineering,* Wiley, New York, 1997.Yen, J., Langari, R., *Fuzzy Logic: Intelligence, Control, and Information,* Prentice Hall, Upper Saddle River, New Jersey, 1999.

Zadeh, L., "Similarity Relations and Fuzzy Orderings," *Information Sciences,* Volume 3, pp. 177–200, Elsevier Science Publishers, 1971.

CHAPTER 4
Learning and Control

In this chapter, decision-making in various applications, including control, pattern recognition, human-machine interfaces, robotics, and apps are examined.

4.1 Architecture of Fuzzy Rule Clusters

In many practical applications, a system of fuzzy *if/then* rules implements a strategy for making decisions dynamically, that is, as a function of time. The decisions can be for the automatic control of machines or devices, but it could also be for decision language-games in business, economics, scientific modeling, social computing, or simply apps for inquiries, preferences, or needs of users in a variety of situations. In this chapter, we will examine dynamic decisions based on control. In Chap. 5, we will examine how this could work on forecasting problems.

Although statistical approaches to learning can be trained on big data volumes, including the massive corpus of text now available online, ever-growing complexity requires strategy over statistical tools. Fuzzy *if/then* rules can encode language-games for decisions pertaining to knowledge verification, data validation, text follow-up, and cross-validation of sources. It is easy, however, to analyze and learn how to design the structure of dynamic decision-making systems by emulating the language-game of fuzzy control.

Fuzzy *if/then* rules associate various situations which we think of as causal conditions and which we place in the *LHS* of a rule, with conclusions or actions to be taken and which we place in the *RHS* of the given rule. As we saw in Chap. 3, fuzzy *if/then* are fuzzy relations which using implication operators (see Table 3.2) and operators for the connective *ELSE* we can agglomerate in rule clusters (*RC*). Each cluster encodes one or more language-games coordinating activities related to a task.

The overall architecture of a fuzzy control system using fuzzy rules is shown in Fig. 4.1. This structure can be seen in numerous

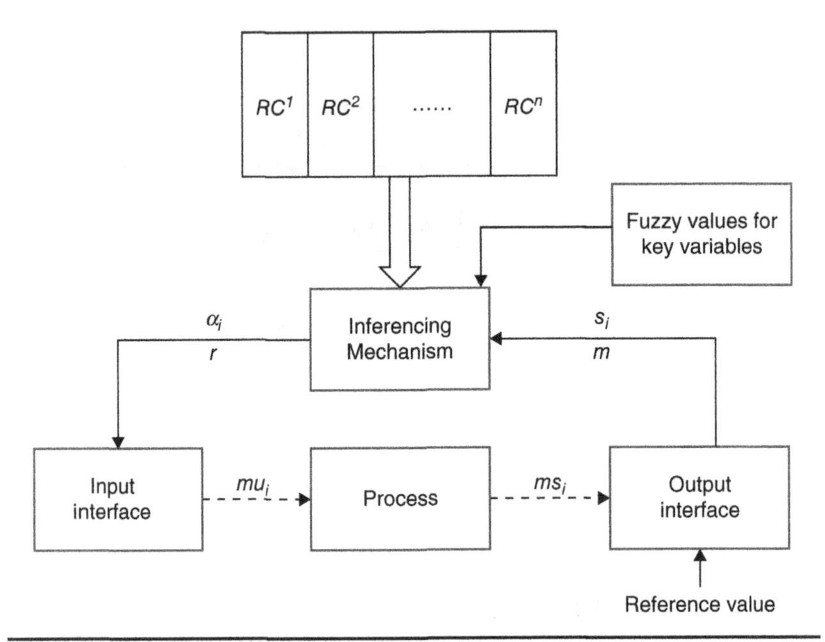

FIGURE 4.1 Architecture of a fuzzy control or dynamic decision-making system.

applications but again, it is not limited to control. For this reason, we call the key component at the center of Fig. 4.1 *Inferencing Mechanism*. In control applications this would simply be called the *Controller*. A great variety of processes, including supply chain, management, educational, and even services in social media or entertainment can be modeled in analogous ways and hence this structure is applicable to any process in need of dynamic decision-making. What gives a fuzzy controller its identity and usefulness is the totality of the rules that are available to decide and act. The rules contain all necessary information for the implementation of a specific control protocol to accomplish goals in a particular process. This is an important point to emphasize because the linguistic nature of this approach is what accounts for its success. The theory of fuzzy logic is just a medium to translate a language-game into rules for making correctible, explainable, and foremost useful decisions dynamically.

The component called **output interface** in Fig. 4.1 catalogues cause-and-effect associations on a given domain and translates these associations into fuzzy sets. In the case of control, the interface may also include analogue to digital converters, scaling coefficients and quantizers, smoothing algorithms, error calculators and rate and corrections of the control or decision-variables and for the values of these variables. The process of translation may involve its own rules, referred to as translation rules, as summarized by Jager and others. In broader decision-making applications, they involve: (a) modification rules, (b) composition rules, (c) quantification rules, and (d) qualification rules (Jager, 1995). These are various setpoints that are involved in the operation of the controller of decision-making algorithm.

The **input interface** on the other hand, transforms the fuzzy output of the fuzzy algorithm (what is also called the resultant membership function) into a crisp (not fuzzy) value that can be used in a process, and thus it involves defuzzifiers, scaling coefficients, integrators of inputs and outputs. It should be noted that this is input to the process. In itself, and as seen in Fig. 4.1, it is the output of the Controller or Inferencing Mechanism.

In Fig. 4.1, a library called "Fuzzy Values for Key Variables" is shown where we store predefined fuzzy sets over appropriate universes of discourse and fuzzy variables that are involved in the calculations of the controllers as well as the universe of discourse found on either the input or output side of the controller. These components can be subjected to machine learning (ML) process (not shown in Fig. 4.1) where the shapes and numbers of fuzzy sets can be modified shifted, scaled, or otherwise modified on the basis of monitoring the performance of the system. A variety of learning tools are now available to facilitate convenient modification of the values in this library.

The *Inferencing Mechanism*, or the Controller, shown in Fig. 4.1 uses the available rules in the *Rule Clusters* above it to draw conclusions based on the current situation, or state, of the system. These decisions are available during the next step in the dynamic process which proceeds in predefined time steps.

Consider an *if/then* rule as a situation-action pair, often denoted simply as $s \rightarrow a$, which symbolically can take the form

$$R^j: \textit{if } x \textit{ is } A^j \textit{ then } y \textit{ is } B^j \tag{4.1}$$

We can write the rule (4.1) in the convenient format of the shorthand $s \rightarrow a$, as

$$R^j: s^j \rightarrow a^j \tag{4.2}$$

where s^j stands for the *LHS* proposition which encodes situation j, and a^j for the *RHS* proposition of the rule encoding action j. This shorthand is convenient for the symbolic and algebraic operations on *if/then* rules and RC. As seen in Fig. 4.1, sensor data or measurements and possibly other sources of data (a variety of data) that collectively characterize the state of a situation are sent as input to the decision-making or "Inference Mechanism."

If we have N rules in a cluster R^N, we can describe them in the convenient shorthand notation as

$$\begin{aligned} R^N &= \{R^1, R^2, \cdots, R^n\} \\ &= \{s^1 \rightarrow a^1, s^2 \rightarrow a^2, \cdots, s^n \rightarrow a^n\} \\ &= \varphi_{\beta}{}_{j=1}^{j=n} (s^j \rightarrow a^j) \end{aligned} \tag{4.3}$$

which contains information pertaining to what actions, a^j, need to be taken for a given set of situations, s^j. Superscripts are used in order to differentiate the rules in a given RC. It should be recalled from Chap. 3 that the operator φ_β refers to how the rules will be joined through an appropriate choice of the connective *ELSE*, which as we have seen in Chap. 3 depends on the implication operator user (see Table 3.2).

As alluded above, the situations s^j and the action a^j can include several variables. For instance, the overall situation s^j can be viewed as the cartesian product of m variables:

$$s^{j1} \times s^{j2} \times \cdots \times s^{jm}$$

and, the action part of a rule, that is, a^j can be considered as the union of r variables:

$$a^{j1} + a^{j2} + \cdots + a^{jr}$$

Hence the totality of the rules can be written as:

$$R^N = \varphi_{\beta}{}_{j=1}^{j=n}\, s^{j1} \times s^{j2} \times \cdots \times s^{jm} \rightarrow a^{j1} + a^{j2} + \cdots + a^{jr} \tag{4.4}$$

Using the property of distributivity, we can rewrite the above equation as:

$$\begin{aligned} R^N &= \{\varphi_{b}{}_{j=1}^{j=n}\, s^{j1} \times s^{j2} \times \cdots \times s^{jm} \rightarrow a^{j1}, \\ &\quad f_{b}{}_{j=1}^{j=n}\, s^{j1} \times s^{j2} \times \cdots \times s^{jm} \rightarrow a^{j2}, \cdots, \\ &\quad \varphi_{b}{}_{j=1}^{j=n}\, s^{j1} \times s^{j2} \times \cdots \times s^{jm} \rightarrow a^{jr}\} \\ &= \bigcup_{p=1}^{r} \varphi_{b}{}_{j=1}^{j=n}\, s^{j1} \times s^{j2} \times \cdots \times s^{jm} \rightarrow a^{jp} \\ &= \bigcup_{p=1}^{r} RC^p \end{aligned} \tag{4.5}$$

Schematically, Eq. (4.5) is illustrated in Fig. 4.2, where r *Rule Clusters* of n rules capture m situations resulting in a total of r actions in a structure that very clearly shows the distributive nature of fuzzy rules.

Figure 4.2 represents an architecture where rules in clusters can manage situations of multiple inputs/outputs. The set R^N includes r subsets of rules (Rule Clusters, RC), one for each consequent or *RHS* variable. Every subset of rules contains n rules in which the antecedent or *LHS* has m variables (causes) and the action part (*RHS*) only one variable. In this way, an *Inference Mechanism* with multiple inputs/outputs

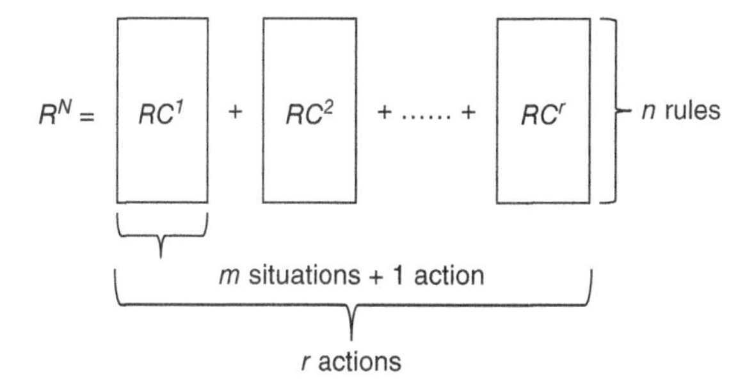

FIGURE 4.2 A fuzzy *if/then* rule architecture where r *rule clusters* (*RC*) of *n* rules each can determine *r* actions.

can be considered or transformed into a number of controllers with multiple inputs (in the *LHS*) and only one output (in the *RHS*).

At any given time $t = i$, an actual *situation/action* of the system is written using subscripts, so that for example the rule

$$s_i \rightarrow a_i$$

describes the state of the system in at the moment in time, $t = i$. The discrete set of time steps $\{i\}$ may be infinite in size (typically very large). It describes the dynamic path of *time trajectory* of the system. The totality of fuzzy rules involved has to infer the action a_1 from the real condition s_1. To achieve this goal *Generalized Modus Ponens (GMP)* is used with composition of fuzzy relations (as seen in Chap. 3). So, for instance, for one rule with one antecedent and one consequent, we have

$$\begin{array}{lll} s^j \rightarrow & a^j: & \text{if } x \text{ is } A \quad \text{then} \quad y \text{ is } B \\ & s_1: & x \text{ is } A' \\ \hline & a_1: & \qquad\qquad\qquad y \text{ is } B' \end{array} \tag{4.6}$$

For *n* rules, we can write *Generalized Modus Ponens* in the following format:

$$\begin{array}{ccc} R^N & \varphi_{\beta}{}_{j=1}^{j=n} \quad s^j \rightarrow a^j & \left[\begin{array}{c} s^j \rightarrow a^j \\ \\ \dfrac{s_1}{a_1^j} \end{array} \right] \\ \dfrac{s_1}{a_1} = & \dfrac{s_1}{a_1} = \varphi_{\beta}{}_{j=1}^{j=n} & \end{array}$$

where,

$$a_i = \varphi_{\beta}{}_{j=1}^{j=n} \quad a_1^j \tag{4.7}$$

The operation in (4.7) represents the matching of the real situation at any given time with fuzzy *if/then* rules capturing generic situation/action pairs.

Applying composition in a sequential manner leads a *Degree of Fulfilment*, DOF_1^j, is obtained for each rule that varies in time. The DOF_1^j is a number between 0 and 1. It measures the extent to which the *LHS* of a rule is matched against an actual situation at time i. This way the action a_1^j is a modified version of a^j, where, in essence, we have,

$$\begin{aligned} &if\ DOF_i^j = 0\ then\ a_i^j = 0\ AND \\ &if\ DOF_i^j = 1\ then\ a_i^j = a^j \end{aligned} \tag{4.8}$$

For a rule-based system of multiple inputs/outputs, with

$$R^N = \bigcup_{p=1}^{r} RC^p$$

the situation described in (4.8) can be extended as follows:

$$\begin{matrix} RC^1 \\ s_i \\ a_{i1} \end{matrix} \quad = \varphi_{\beta}{}_{j=1}^{j=n} \left[\begin{array}{c} s^{j1} \times s^{j2} \times \cdots \times s^{jm} \to a^{j1} \\ \dfrac{s_{i1} \times s_{i2} \times \cdots \times s_{im}}{a_{i1}^{j1}} \end{array} \right]$$

$$\begin{matrix} RC^2 \\ s_i \\ a_{i2} \end{matrix} \quad = \varphi_{\beta}{}_{j=1}^{j=n} \left[\begin{array}{c} s^{j1} \times s^{j2} \times \cdots \times s^{jm} \to a^{j1} \\ \dfrac{s_{i1} \times s_{i2} \times \cdots \times s_{im}}{a_{i2}^{j2}} \end{array} \right]$$

....

$$\begin{matrix} RC^r \\ s_1 \\ a_{ir} \end{matrix} \quad = \varphi_{\beta}{}_{j=1}^{j=n} \left[\begin{array}{c} s^{j1} \times s^{j2} \times \cdots \times s^{jm} \to a^{jr} \\ \dfrac{s_{i1} \times s_{i2} \times \cdots \times s_{im}}{a_{ir}^{jr}} \end{array} \right] \tag{4.9}$$

where in Eq. (4.9), we have:

$$a_{i1} = \varphi_{\beta}{}_{j=1}^{j=n}\ a_{i1}^{j1}$$

$$a_{i2} = \varphi_{\beta}{}_{j=1}^{j=n}\ a_{i2}^{j2}$$

....

$$a_{ir} = \varphi_{\beta}{}_{j=1}^{j=n}\ a_{ir}^{jr} \tag{4.10}$$

Thus, for a specific composite situation $s_{i1} \times s_{i2} \times \cdots \times s_{im}$ and a specific rule j, the degree of fulfillment DOF_i^j will be the same for all the subsets of rules in a rule cluster RC and hence it only needs to be computed once for all rules as shown in (4.10). In this way the degree of fulfillment DOF_i^j can modify all the consequents of the rules, that is, $a^{j1}, a^{j2}, \ldots, a^{jr}$ that contribute to all final actions $a_{i1}, a_{i2}, \ldots, a_{ir}$ at the time $t = i$.

The rule clusters RC provide several parameters that can be subjected to machine learning and thus lead to language-games modifiable on the face of experience. The parameters include, but are not limited to, quantities that describe the membership functions, such as the edges of triangular or trapezoidal numbers, quantities that describe bell-shaped memberships, but also rule weights or parameters involved in Sugeno-type rules (Terano, 1989), or degrees of fulfillment for rules, as in the discussion above.

4.2 Constructing a Rule-Based Controller and Interfacing It with Its Environment

The variables of the Inferencing Mechanism in Fig. 4.1 can be classified either as input variables which describe the state of affairs or situation of the system in a generic sense, or as output variables that constitute signals for actions sent to actuators or operators. In actual control situations, the variables closer to classical controllers focus on the discrepancy between a reference value, or setpoint, called the *error* and variables based on the error, such as the *derivative of error*, which is the difference between the error in two adjacent time steps, or the *integral of errors* over a predefined window of time. The last variable could signify the history of errors and it is variable that carries some inertia; in other words it acts to smooth the control action, whereas the derivative may act fast and it is often seen as a variable that anticipates the trajectory of the system. The control strategy is to take action proportional to the error, the derivative of error or the integral of error which are the principal modalities of action found in the proportional-integral-derivative (PID) controller. Let us next look at these variables in some detail.

Input Variables

As in a conventional controller, a fuzzy controller using linguistic variables may have as input variables the *error*, formed between a measurement and a reference value S as well as the *rate of change of error* or the *sum of errors*. The first can be quantitatively defined as

$$\langle s_i \rangle = S - ms_i$$

where S is the setpoint (reference) value, ms_i the measurement at time i, and s_i the error.

The *rate of change of error*, that is, the first derivative of the error, can be defined as:

$$\langle s_i' \rangle = \langle s_i \rangle - \langle s_{i-1} \rangle = ms_{i-1} - ms_i$$

The *sum of errors* (that is the "history" or integral if we had continuous signals) is defined as:

$$\langle \overline{s}_i \rangle = \sum_{k=0}^{i} ms_k$$

We can also define the rate of the *rate of change of the error*, that is, the second derivative of the error, as:

$$\langle s_i'' \rangle = \langle s_i' \rangle - \langle s_{i-1}' \rangle = -\, ms_i + 2\, ms_{i-1} + ms_{i-2}$$

Output Variables

The linguistic variables of the output of the controller may be based in the normal values of a variable, mu_i, or in the difference or step-change of the variable with respect to its value in the previous time step, Δmu_i, in which case we need an external summation or integration to transform it to normal values, that is,

$$mu_i = mu_{i-1} + \Delta mu_i$$

In the last case (that is, the equation above), we have a more efficient implementation, since it requires a smaller number of data in the universe of discourse of the output signal in order for the Inference Mechanism to provide output of reasonable accuracy.

Smoothing, Scaling, Quantization

Smoothing of the values of the variables may be used in order to avoid rapid changes during sampling that distract from the general trend of a variable. A smoothing function often used in control systems is

$$\begin{aligned} \langle\langle s_i \rangle\rangle &= E(\langle s_i \rangle) \\ &= 0.9\langle\langle s_{i-1} \rangle\rangle - 0.1\langle s_i \rangle \end{aligned}$$

Scaling may be needed in order to transform input data from sensors as well as from other sources into predefined ranges, or, in order to transform these values into data that corresponds to physical quantities.

Quantization of the values of the variables may be used when we work with discrete universes of discourse. The quantization function $QL(\cdot)$ takes the form

$$\begin{aligned} s_i &= QL(\langle\langle s_i \rangle\rangle) \\ &= \frac{q}{k} \;\; if \;\; \frac{(q-9.5)}{k} < \langle\langle s_i \rangle\rangle \leq \frac{(q+9.5)}{k} \end{aligned}$$

$$\text{for} \quad q = 0, \;\; \pm 1, \;\; \pm 2, \ldots, \pm \frac{(-1)}{2}$$

where L is the number of elements in the universe of discourse and k determines the step of the scale, with saturation of values beyond $\frac{(L-1)}{2}$. The tolerance band for zero error is $\pm\frac{1}{2k}$. If we define our quanta based on a zero-error tolerance band, we may have saturation in the function $QL(\cdot)$ and beyond this limit, all measurements may fall in the category *LARGE*.

Alternatively, the quantization function may be defined with reference to a large number of elements in the universe of discourse, the smallest of which can be defined from the zero-tolerance band. In this way we may avoid the discontinuity introduced by this method of quantization. An alternative is to use a universe of discourse with logarithmic scale, or some logarithmic quantization method which would provide sufficient precision for small values in the variables while at the same time include all the values in the range of measurements with a relatively small number of quanta.

4.3 Defuzzification Methods

A fuzzy control algorithm outputs the composite of several rules in a resultant fuzzy output $\mu_{OUT}(u)$. The actuators of an automatic control system do not take fuzzy values as input. Hence, we need to select a crisp number u^* representative of $\mu_{OUT}(u)$. This is a process known as *defuzzification*. The most frequently used defuzzification methods are the *centroid* or *center of area* (*COA*), the *center of sums* (*COS*) and *mean of maxima* (*MOM*).

Center of Area (COA) Defuzzification

In *COA* defuzzification, the crisp value u^* is taken to be the geometrical center of the output fuzzy value $\mu_{OUT}(u)$ where $\mu_{OUT}(u)$ is formed by taking the union of all the contributions of rules whose $DOF > 0$. The center is the point which splits the area under the $\mu_{OUT}(u)$ curve in two equal parts. The defuzzified output for a discrete universe of discourse is defined as

$$u^* = \frac{\sum_{i=1}^{N} u_i \, \mu_{OUT}(u_i)}{\sum_{i=1}^{N} \mu_{OUT}(u_i)} \tag{4.11}$$

where the summation (integration) is carried over (discrete) values of the universe of discourse u_i sampled at N points. *COA* is the most often used defuzzification method.

Center of Sums (COS) Defuzzification

To address some problems associated with *COA* and take into account the overlapping areas of multiple rules more than once, a variant of *COA* called *Center of Sums* (*COS*) is used. *COS* builds the resultant membership function by taking the sum (not just the union) of output from each contributing rule. Hence overlapping areas are counted more than once. *COS* is actually the most commonly used defuzzification method. It can be implemented easily and leads to rather fast inference cycles. It is given by

$$u^* = \frac{\sum_{i=1}^{N} u_i \cdot \sum_{k=1}^{n} \mu_{B'_k}(u_i)}{\sum_{i=1}^{N} \sum_{k=1}^{n} \mu_{B'_k}(u_i)} \tag{4.12}$$

where $\mu_{B'k}(u_i)$ is the membership function (at point u_i of the universe of discourse) resulting from the firing of the kth rule.

Mean of Maxima (MOM) Defuzzification

One simple way to defuzzify the output is to take the crisp value with the highest degree of membership in $\mu_{OUT}(u)$. Oftentimes though there may be more than one element in the universe of discourse having the maximum value. In such cases we may randomly select one of them or, even better, take the mean value of the maxima. Suppose that we have M such maxima in a discrete universe of discourse. The crisp output can be obtained by

$$u^* = \sum_{m=1}^{M} \frac{u_m}{M} \tag{4.13}$$

where u_m is the mth element in the universe of discourse where the membership function of $\mu_{OUT}(u)$ is at the maximum value, and M is the total number of such elements.

MOM defuzzification is faster than *COA* and allows the controller to reach values near the edges of the universe of discourse. A disadvantage of this method, however, is that it does not consider the overall shape of the fuzzy output $\mu_{OUT}(u)$ (Tsoukalas, 1997).

Example 4.1 Fuzzy Control System for Water Tank Level A fuzzy logic system is used to control level in a tank through valve, as shown in Fig. 4.3, where a tank of height 5 m and diameter 2 m has a steady flow of 0.4 m^3/min filling the tank while an outlet flow of capacity 1 m^3/min with a variable valve is emptying the tank. Level control problems are found in many diverse applications (Graham, 1988; Hirota, 1993; Lee, 1980; Sugeno, 1984). The valve capacity ranges from fully closed (0.0) to fully open (1.0), that is, 0% to 100%. The setpoint level of the liquid in the tank is 2.5 m.

A *Rule Cluster* with nine fuzzy *if/then* rules is used. It is implemented in the Python language, using the "skfuzzy" library (the code for the RC is given in Appendix 1). The physical process of this system is the liquid problem as described. The method of inference used is the *Generalized Modus Ponens*. The fuzzy relation model of the rules is obtained with *Mamdani implication*. The system consists of two inputs, the *height level of fluid in the tank* and the *height level rate change per second (dh/dt)*, and one output, the *output valve opening*. In this example, the linguistic variables are:

height HIGH	*dh/dt POSITIVE*
height AT_SET_POINT	*dh/dt ZERO*
height LOW	*dh/dt NEGATIVE*
valve MUCH_MORE_CLOSED	
valve MORE_CLOSED	
valve NO_CHANGE	
valve MORE_OPEN	
valve MUCH_MORE_OPEN	

where *height*, for example, is a universe of discourse and *HIGH* is a *height* value. It is then possible to write control rules such as

if (height is HIGH) AND (dh/dt is POSITIVE) then
(valve is MUCH_MORE_OPEN)

For each fuzzy subset, there is a membership function which indicates for each item in the universe of discourse its degree of membership. For a universe of discourse *U* having a fuzzy subset *A*, the membership of some object *x* in *A* is denoted as follows:

$$\mu_A(x)$$

In the situation where two linguistic variables are related by logical *AND*, it is necessary to define the membership function of the conjunction *P AND Q* given the membership functions for each of *P* and *Q*. This membership function is defined as:

$$\mu_{P\times Q}(p, q) = \wedge [\mu_P(p), \mu_Q(q)]$$

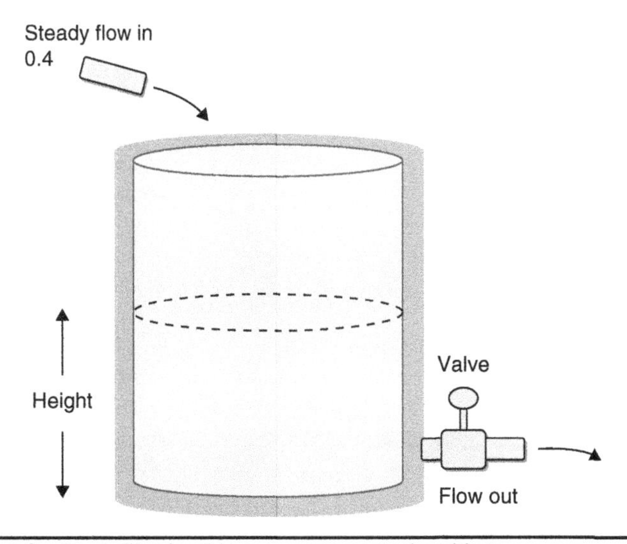

FIGURE 4.3 Controlling level is a common control problem.

Given an implication of the form *if x* is *A then y* is *B*, the membership function of *B* given *A* is stated as:

$$\mu_{B:A}(u,v) = \wedge[1, (\mu_A(u) + \mu_B(v))]$$

Generalized Modus Ponens in fuzzy logic takes the following form:

$$\frac{\begin{array}{c} \textit{if x is A then y is B} \\ \textit{x is A'} \end{array}}{\textit{y is B'}}$$

The grade of membership of *y* can be obtained from composition, that is, using the equation:

$$\mu_y(v) = \vee[\wedge[\mu_A(u), \mu_{B:A}(u,v)]]$$

Through a combination of conjunction and *GMP* it is possible to write and evaluate fuzzy *if/then* rules as examined earlier in this chapter. In a rule-based fuzzy logic system, all rules are fired and the final conclusion may be thought of as having contributions from all rules. However, the strength of contribution of a rule is determined by the conjunction of the antecedents. The membership function of the consequent is then scaled to reflect this strength. In this system, the rule base is formed as:

if height is HIGH	*AND dh/dt is POSITIVE*	*then value is MUCH_MORE_OPEN*
ELSE		
if height is HIGH	*AND dh/dt is ZERO*	*then value is MUCH_MORE_OPEN*
ELSE		
if height is HIGH	*AND dh/dt is NEGATIVE*	*then value is MORE_OPEN*
ELSE		
if height is AT_SET_POINT	*AND dh/dt is POSITIVE*	*then value is MORE_OPEN*
ELSE		
if height is AT_SET_POINT	*AND dh/dt is ZERO*	*then value is NO_CHANGE*
ELSE		
if height is AT_SET_POINT	*AND dh/dt is NEGATIVE*	*then value is MORE_CLOSED*
ELSE		
if height is LOW	*AND dh/dt is POSITIVE*	*then value is MORE_CLOSED*
ELSE		
if height is LOW	*AND dh/dt is ZERO*	*then value is MUCH_MORE_CLOSED*
ELSE		
if height is LOW	*AND dh/dt is NEGATIVE*	*then value is MUCH_MORE_CLOSED*

The membership function of the two inputs and one output is shown in Fig. 4.4. In this fuzzy logic system, the input is in the form of numbers and the output must be in the form of one value for each output variable. It is therefore necessary

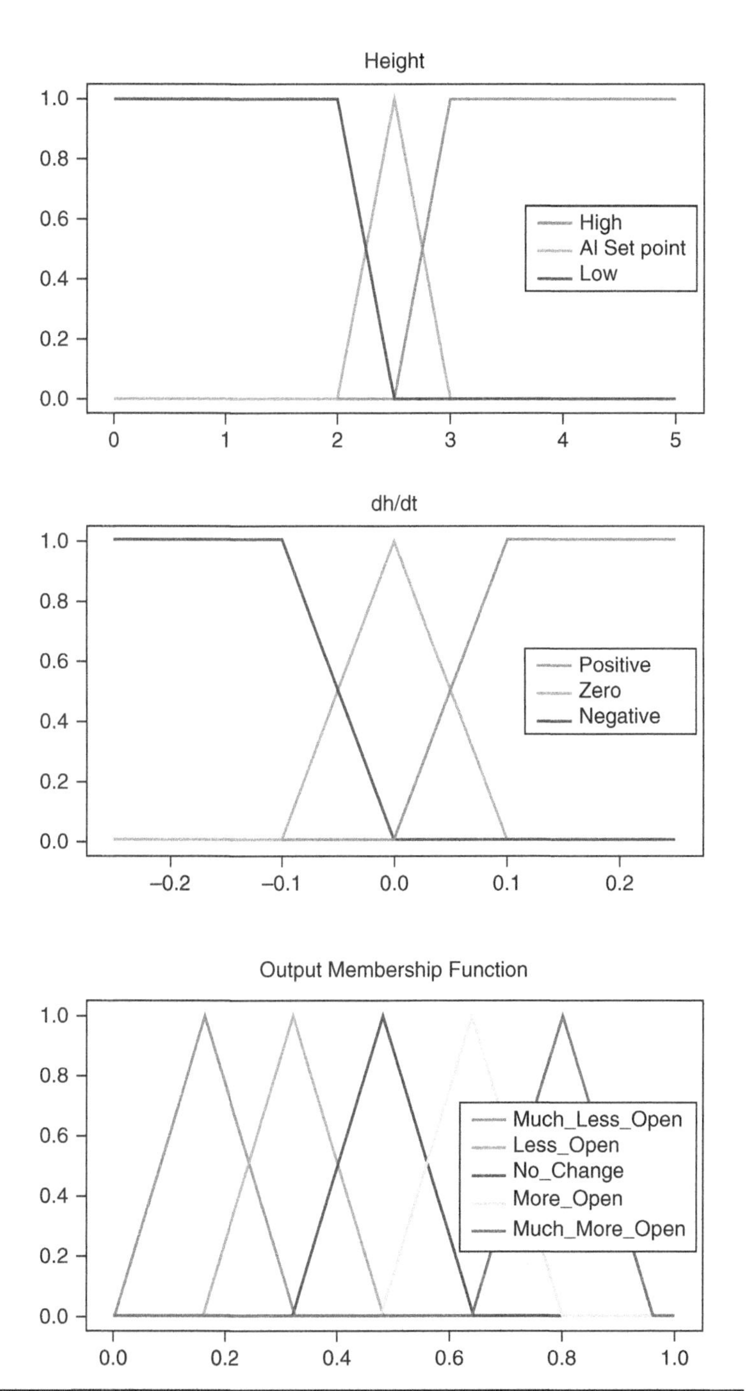

FIGURE 4.4 Membership functions for the values of the variables in *LHS* and *RHS* of the control algorithm.

to convert input numerical values into values for linguistic variables, a process referred to as "fuzzification." Fuzzification can be commonly performed since each input linguistic variable has a membership function attached to it. A challenge arises when we need to extract a number from the consequent membership function, that is, defuzzification. The method employed in this system is the *Center of Area* (*COA*) as described in Sec. 4.3. Given a membership function, the defuzzified value is that value at the centroid of the area under the membership function curve.

The basic operation of the rule-based fuzzy logic system is the inference of the valve opening used to control the flow out of a water tank. To demonstrate an example, we set the height level of fluid in the tank at 2 m, and the height level rate change per second, *dh/dt*, at +0.03. A graphical illustration of how a rule would be evaluated is shown in Fig. 4.5.

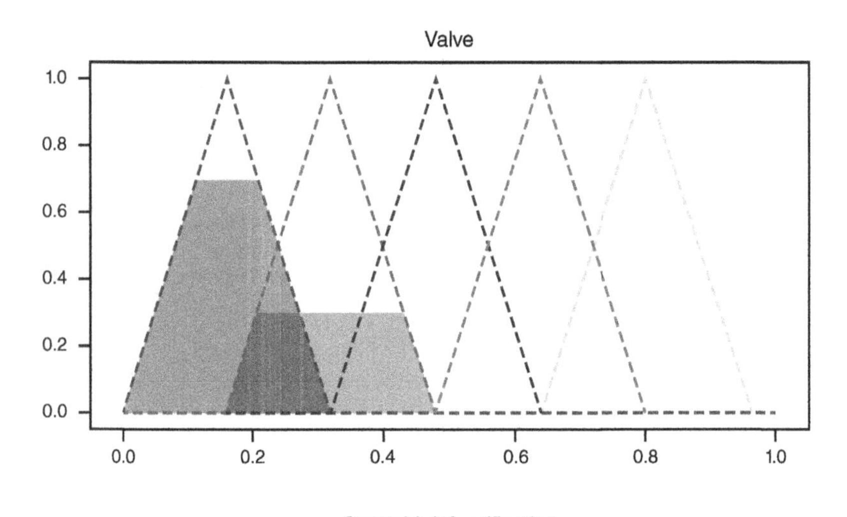

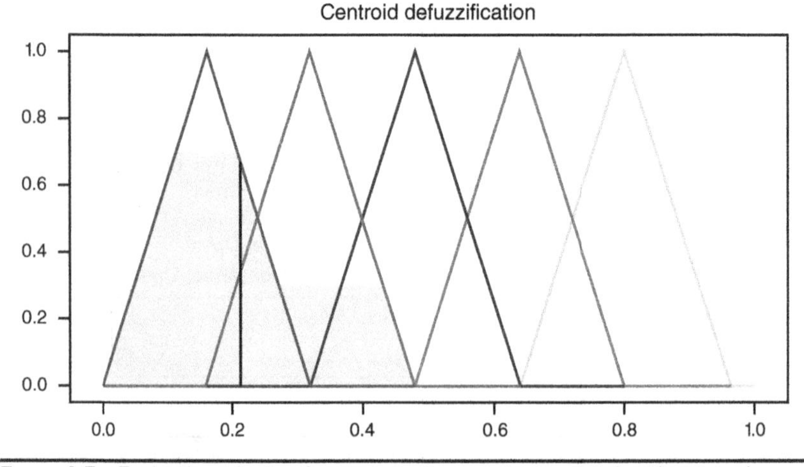

FIGURE 4.5 Evaluation of rules for output valve opening, and defuzzification of results for the output of the valve.

References

Dubois, D., and Prade, H., *Fuzzy Sets and Systems: Theory and Applications,* Academic Press, Boston, 1980.

Graham, B. P., and Newell, R. B., "Fuzzy Identification and Control of a Liquid Level Rig," *Fuzzy Sets and Systems,* Vol. 26, pp. 255–273, 1988.

Hirota, K., *Industrial Applications of Fuzzy Control,* Springer-Verlag, Tokyo, 1993.

Jager, R., *Fuzzy Logic in Control,* Unpublished Ph.D. Dissertation, University of Delft, The Netherlands, 1995.

Mamdani, E. H., "Advances in the Linguistic Synthesis of Fuzzy Controllers," *International Journal of Man-Machine Studies,* Vol. 8, pp. 669–678, 1976.

Sugeno, M., "An Introductory Survey of Fuzzy Control," *Information Sciences,* Vol. 36, pp. 59–83, 1985.

Terano, T., Asai, K., and Sugeno, M., *Fuzzy Systems Theory and Its Applications,* Academic Press, Boston, 1992.

Tsoukalas, L., and Uhrig, R. E., *Fuzzy and Neural Approaches in Engineering,* Wiley, New York, 1997.

CHAPTER 5

Forecasting with Fuzzy Algorithms

This chapter shows how a forecasting process can be implemented very easily through a language-game for control and decision-making including fuzzy and a probabilistic approach.

5.1 Forecasting

Forecasting is a great contributor to intelligence. Predicting variables, anticipating situations, and estimating future developments allow us to steer systems, allocate resources, or even mitigate contingencies that may emerge in the future. Especially for complex networks such as the internet or the power system, the capacity to anticipate may address all kinds of difficult problems, such as a lack of storage and bandwidth management. For instance, the power system has minimal capacity for storage, a resource which is essential in addressing the stochasticity of intermittent power sources. Can this be offset, at least to a degree, by anticipating better future states of the system and thus affecting demand and generation to ensure a balance in supply and demand? The answer to this and other questions depends on our ability to forecast. In this chapter, we focus on demand forecasting as an emblematic forecasting problem but also because of its significance in the evolution of a power system expected to accommodate the electrification of transportation. The essence of this perspective is that forecasting may transform the storage problem into a scheduling problem, the later being easier to exploit since our computational capacities are growing a lot faster than innovations in energy storage.

Forecasting of power demand, commonly called "load," can be broadly classified into three categories due to their objectives and functions: very short-term load forecasting, short-term load forecasting, and, long-term load forecasting.

Very short-term load forecasting is to predict the load 20 to 30 minutes ahead of real time in intervals of 1 or 2 minutes. This load prediction will be dynamically dispatched to each available generator over a moving window of time. Thus, generation control will be accomplished based primarily on anticipated values rather than after-the-fact response. In addition, very short-term load forecasts can also be used to solve the problem of forecasting the load with better exploitation of power tariffs, especially for reducing contractual power to avoid strong penalties. In spite of significant research efforts to improve forecasting accuracy, most progress has been seen in hourly data or daily data. Some researchers use radial basis neural networks, which have fast convergence, to perform this type of forecasting (Verona, 1998). Others use fuzzy logic and autoregressive methods, with results indicating that neural computing and fuzzy logic methods are good candidates for forecasting transient-type problems (that is, when new patterns emerge away from steady state) (Wang, 2002). Support vector machines for short-term load forecasting have been used with noticeable successes (Bougaev, 2002; Bougaev, 2007).

The goals of load forecasting as articulated by Gross are in a broad sense the following (Gross, 1987):

- Daily peak system load
- The values of system load at certain times of the day
- The hourly values of system energy
- The daily and weekly system energy

There are large economic benefits to forecasting improvements due to the fact that the power system is a peak-demand constrained system. In other words, the system operations must be able to satisfy maximum demand, whenever it appears during a 24-hour horizon, to ensure the stability of the system. According to an old but valid assessment about $0.5 M/year per utility is estimated to be saved due to 1% reduction in forecast error (Hobbs, 1998).

Occasionally *very short-term* and *short-term* are lumped together as *short-term forecasting* problems. However, because of the different objectives of these two problems, the input data used and the predicting strategies may be quite different, and it is better to deal with the two problems separately.

In order to match the power demand with supply, long-term prediction is also very important because electric utilities expect to have adequate and appropriate generating and transmission capacity resources for several years ahead. In predicting long-term load, the weather factors are not important anymore because weather prediction is highly unreliable beyond a 2-week horizon. Instead, economic factors such as gross domestic product (GDP), population, and economic growth may dominate future demand, and hence affect planned capacities of generation. Yet some of the strategies used in short-term load forecasting can also be applied in long-term load forecasting.

Conventional statistical load forecasting methods use time series and auto-regressive models. In time series models, the data of the previous hours or days are utilized by extrapolation. It is a very simple model. No complicated summarization or derivation is needed. But the disadvantage is that this technique is susceptible to errors under conditions of sudden and large weather variations since these effects are not modeled.

In auto-regressive models, more information or factors influencing the load value are taken into account. Large amounts of data are analyzed and summarized. The load is assumed to include two parts: *a slowly changing part* and *a fast changing part*. The first component can be represented by a periodic time function assuming normal weather condition. The residual component is weather-sensitive, which is relatively small in quantity compared with the first one. This technique needs more information than time series models but results on more accurate results. However, analyzing large amounts of data and accurately approximating the whole system are too difficult to achieve. Typically, in a short-term load prediction model, temperature and previous load information are utilized to predict future values. For different times during the day, different correlations are applied. In addition, the correlations are updated with incoming new data. This model is appropriate for load with smaller oscillation (Basu, 1992).

Alternatively, the load model may be composed of three parts: the nominal load, the type load, and the residual load (Park, 1991). The nominal load is modeled with Kalman filters, and the parameters of the model are adapted by an exponentially weighted recursive least squares method. The type load component is extracted for weekend load prediction and updated by an exponential smoothing method. The residual load is predicted by the autoregressive model and the parameters of the model are estimated using the recursive least squares method. The results show the error is much smaller than that in literature. Conventional load forecasting is still useful but it is replaced slowly by developing logical and statistical learning approaches with conventional approaches still useful for cross-checking the new approaches.

Synergisms of neural networks with fuzzy logic have gained considerable attention (Tsoukalas, 1997; Wang, 2002). Neural networks consist of a large number of simple, highly interconnected processing elements in an architecture inspired by the structure of the cerebral cortex of the brain. For neural networks, the most attracting property is the self-learning algorithm. Because of this advantage, neural networks do not depend on any functional form of the forecasting model, which allows its adaptivity to changing situations in practice. It is also a good approximator of the nonlinear systems, which have many inputs and outputs and have no explicit mathematical expression. Many different types of neural networks have been proposed to perform load forecasting. The results show the accuracy is very encouraging. The relative error is normally below 2%, which is much lower than conventional methods.

Early neural approaches focused on the embedded periodicity in the load trend. The autocorrelation, partial autocorrelation, and the cross correlation of the loads were used to analyze the load with results showing that the hourly load data from today, yesterday, and the same 2 days of the previous week may be useful (Chen, 1992). In addition, the load is strongly correlated with the day of the week (weekend or not), the time of the day (morning or afternoon), the value of the temperature. Hourly temperature of the forecast day, average hourly load values obtained by averaging the previous three weeks' load data for the same day type, average load value for the previous hour are utilized as input data. Although the high accuracy obtained by researchers, some persisting problems are:

- The load in weekend and in holidays is difficult to predict. Some of the researchers simply treated them as bad data or regarded their value as half of the normal value.[16]
- The load change for unexpected weather change or specific event is not considered by most of the researchers.
- Some of the input data such as temperature and humidity are not accurate enough, which may impact the final results.

Fuzzy logic is effective when it is used to handle the following problems: uncertainties in input data, properties hidden in large amounts of data, utilizing qualitative information. In load forecasting problems, fuzzy logic can be useful in many aspects. One common approach is to utilize fuzzy *if/then* rules to capture uncertainties in weather variables and statistical models. Experienced operator heuristic rules are imbedded in the knowledge base.

One of the most important properties of fuzzy logic is its ability to handle uncertainty in input. Instead of using temperature and humidity values, the researchers used a parameter to fuse the information in these parameters named discomfortable index (DI). It is reasonable to utilize this parameter if we realize:

- People do turn on or off their air conditioners when they do not feel good.
- The predicted weather values are highly unreliable.

Another important advantage of using DI is the accumulative effect can be taken into consideration. The relationship between DIs on the next and previous days are investigated in the paper. The power operators pointed out that there exists the correlation between them. An explicit mathematical expression is given in this paper based on the dry-bulb, wet-bulb temperature, and the correlation (Mori, 1999).

Although fuzzy forecasters can provide roughly the same accuracy as neural networks, they are short of ability of self-learning. The expert systems cannot adapt themselves to new or changing environment. Therefore, they cannot handle sudden changes in load.

In order to enable the forecaster to think and study as humans, researchers have been trying to incorporate neural networks and fuzzy logic systems. It seems to be the most promising field of load forecasting (Bakirtzis, 1995).

There are generally two kinds of neuro-fuzzy systems. The first one is a fuzzy expert system using ANN structure and back propagation strategy. In this methodology, fuzzy numbers replace the neurons and the weights are replaced by fuzzy rules. The second-class neural fuzzy systems utilize normal neural networks and a pre or post fuzzy logic module to correct the provisional result provided by the ANN. Normally, the favorite ANN structures are feed-forward back-propagation, recursive and radial basis neural network. Some of the research has shown the structure of the neural network does not influence the results.

Bakirtzis developed a fuzzy-neural network (FNN) for forecasting where a fuzzy system has the network structure and the training procedure of a neural network. The FNN initially creates a rule base from existing historical load data. The parameters of the membership functions are adjusted and tuned according to back propagation strategy. One advantage is the training process is much faster than that of an ANN. Another one is qualitative information like explicit logic is applied in the network, which is easier to be understood.

In neurofuzzy systems, the non–weather-sensitive component can be simulated by neural networks while the pure weather-sensitive component is processed with a fuzzy logic. For the non–weather-sensitive component, the input data include time of the day, the past 2 hours' load information and the load information of the same day in last week (Tsoukalas, 1997).

Another application of fuzzy logic is preprocessing of the data. The data, especially the weather factors, are fuzzified and classified by preprocessing modules, based on the temperature and the humidity. Then the fuzzified weather conditions are fed with other data such as day of the week, hour of the day, and previous load into a neural network. The preprocessor fuzzifies the input data of temperature, time index, and previous load and load change rate.

5.2 Fuzzy Logic Forecasting

Short-term demand forecasting plays a major role in ensuring a secure and cost-effective operation of power systems, such as nuclear power plants. This type of forecasting can provide an accurate, fast, and robust forecasting, which is important for the electric utility and consumers. Short-term forecasting is needed to supply necessary information for the system management of day-to-day operation. Hence, this example is focused on short-term load forecasting.

Since (i) a power system is affected by different factors (time of the day, weather conditions, electricity price, etc.) and is characterized by nonlinear and uncertain conditions, and (ii) the relationship between load demand and the factors affecting it is complex and thus statistical models may fail to fit the load curve, a language-game approach is adopted to predict the short-term power demand, as in the work presented in Al-Anbuky (1995).

The accuracy of a prediction model is based not only on the methodology used but also on the selection of inputs. Figure 5.1 shows both the inputs and the steps conducted to estimate the output of the fuzzy model, which is the predicted electricity demand.

As we can observe in Fig. 5.1, the input variables inserted into the model are fuzzified in such a way as to extract the prediction system's *if/then* rules. The definition of the membership function for each variable is a key element of this step of the process. In this application, we use the triangular and trapezoidal membership functions. In general, triangular and trapezoidal membership function can provide flexibility in data handling.

The fuzzy sets and membership functions for every variable of the application are shown in Figs. 5.2–5.5. We see in these figures that each fuzzy variable of the system is decomposed into a reasonable number of fuzzy regions following the rule of thumb of choosing an odd number of labels associated with a variable (Cox, 1992).

More specifically, the fuzzy variable *time of day* is categorized into three fuzzy values as follows:

1. Time periods that pertain to hours of sleep
2. Time periods of work, and
3. Time periods of leisure

Similarly, the fuzzy variable of *temperature* takes five values; that is, we have five membership functions, describing *COLD*, *COOL*, *COMFORTABLE*, *WARM*, and *HOT* conditions.

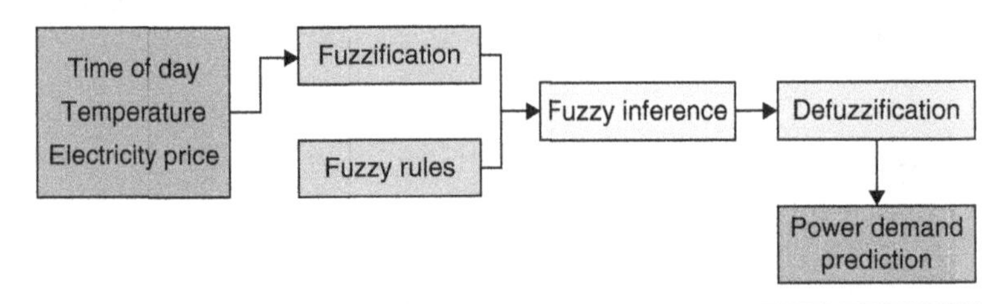

Figure 5.1 A model for power demand prediction.

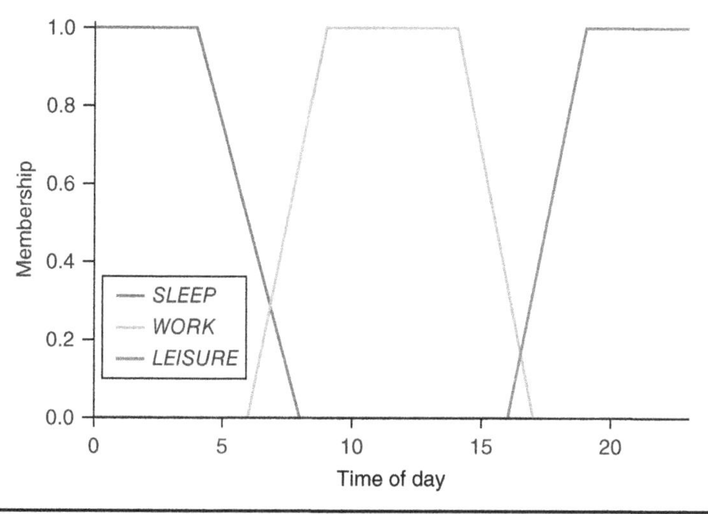

FIGURE 5.2 Membership functions of the input fuzzy variable time of day.

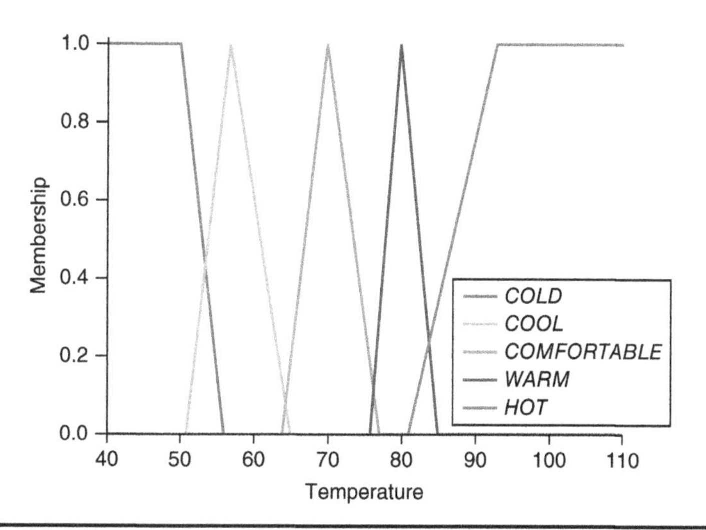

FIGURE 5.3 Membership functions of the values of the fuzzy variable temperature.

The fuzzy variable *price of electricity* has three membership functions, one for *OFF-PEAK*, one for *NORMAL*, and one for *PEAK* charges. Finally, *electricity demand* can be categorized as *LOW*, *MEDIUM*, and *HIGH*.

Determining the fuzzy rules is the next important step. Each fuzzy rule will lead to a weighted fuzzy output variable. Although

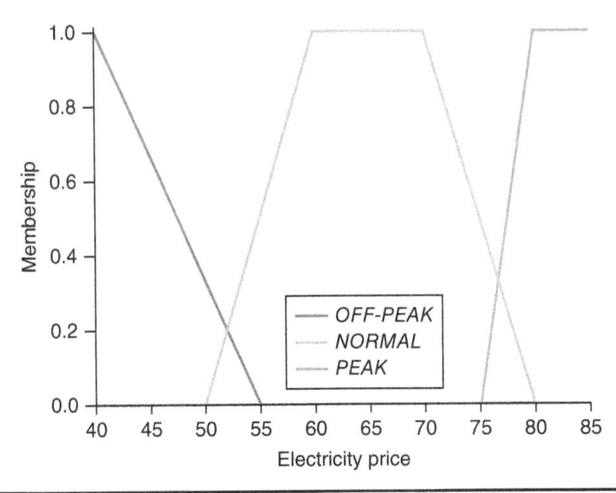

FIGURE 5.4 Membership functions of the input variable electricity price.

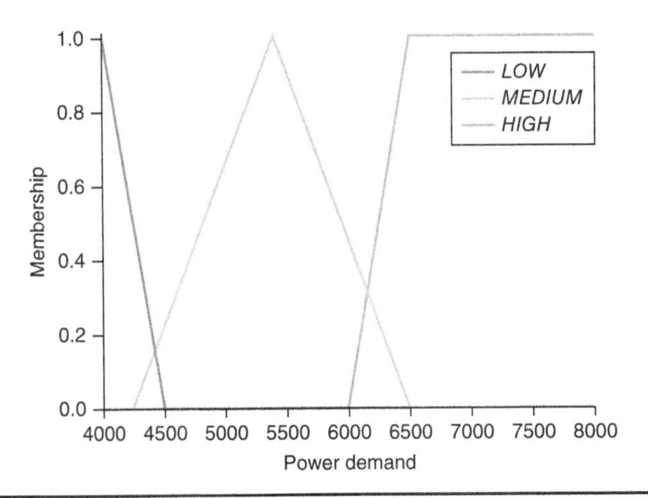

FIGURE 5.5 Membership functions modeling the values of the output variable power demand.

we need all the reasonable possible combinations of the input variables, in this example, we only show three rules. The *if/then* rules are:

1. *If (time of day is WORK) AND (temperature is COLD) AND (electricity price is NORMAL) then (demand is HIGH) ELSE*
2. *If (time of day is SLEEP) AND (temperature is COOL) AND (electricity price is OFF-PEAK) then (demand is LOW) ELSE*
3. *If (time of day is LEISURE) AND (temperature is COMFORTABLE) AND (electricity price is NORMAL) then (demand is MEDIUM)*

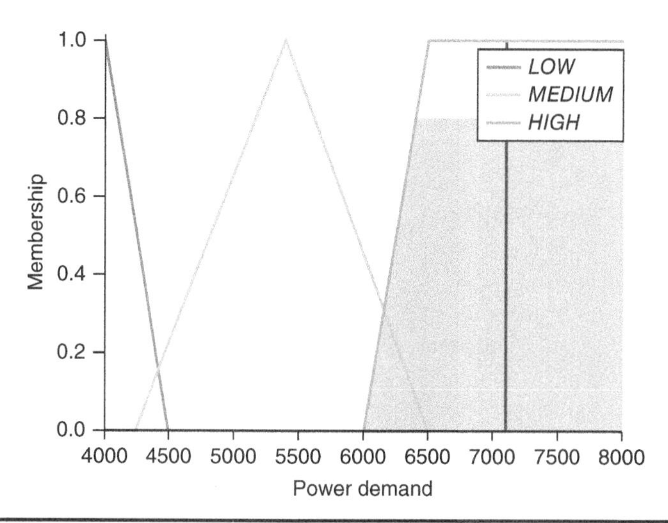

FIGURE 5.6 Output of the fuzzy controller predicting power demand *7096.3* MW when given as input: *time of day = 13, temperature = 45, and electricity price = 58.*

The summation of all the fuzzy outputs based on the fuzzy rules is then defuzzified to produce the power demand. The centroid method for defuzzification is used as the default defuzzification method in the package. This method is a well-known and often used defuzzification method (Tsoukalas, 1997).

As a concrete example, we can insert the following values for the input variables in the fuzzy controller we created: *time of day* = *13*, *temperature* = *45*, and *electricity price* = *58*. The output of the controller for the specific values after defuzzification is equal to *7096.3* MW and is shown in Fig. 5.6. For the specific example, the scikit-fuzzy Python package was used and the code for the implementation of the specific problem is provided in Appendix B.

5.3 Predicting Demand with a Probability Model

In order to contrast with the fuzzy approach, we build a simple probabilistic model, according to which the demand for the following day is predicted based on the demand for the current day. This will produce a discrete evolution in time, showing the state of the demand each day. This model is deterministic in the sense that knowledge for the current demand will lead to knowledge of a future state.

The demand can have three states—high, medium, and low—which are shown in Fig. 5.7. We assign probabilities connecting all the states. For example, if the current demand is high, there is a 0.5 probability that it will also be high for the following day, a 0.25 probability that it is going to be medium, and a 0.25 probability that it is going

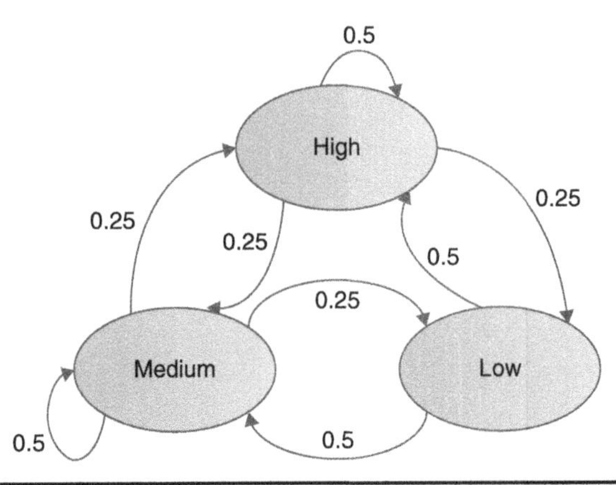

Figure 5.7 A transition diagram model for demand prediction.

to be low. These probabilities can be deduced from empirical knowledge, or by taking into account knowledge about the demand for the same day in previous years.

The system is dynamical, meaning that it changes with time. Equation (5.1) represents the state vector for the current demand. For example, if the current demand is medium, the vector will be of the form [0 1 0]. Equation (5.2) predicts the demand for the following day by multiplying the matrix A containing the transition probabilities with the demand for the current day.

$$x_{today} = \begin{bmatrix} pr(H) \\ pr(M) \\ pr(L) \end{bmatrix} \tag{5.1}$$

$$x_{tomorrow} = A \cdot x_{today} \tag{5.2}$$

The following is a simple code implemented in Python based on the procedure described above. We first import the necessary packages:

```
import numpy as np

from matplotlib import pyplot as plt
```

Next, we define the matrix A, as well as the vector xtoday representing the current demand. According to Eq. (5.2), we then predict the demand for the 50 days that follow and store the values in an array.

```
A = np.array([[0.5,0.25,0.5], [0.25, 0.5, 0.5],
[0.25, 0.25, 0]]);
```

```
xtoday = np.array([[1],[0],[0]]);
demand = np.zeros((3,50));
demand[:,0] = xtoday[:,0];
for k in range(50):
    xtomorrow = A@xtoday;
    demand[:,k] = xtomorrow[:,0];
    print(k);
    print(xtomorrow);
    xtoday = xtomorrow;
```

Lastly, we plot the results, which are shown in Fig. 5.8. We can see that, after only a few days, the probabilities converge to their final values and the system reaches a steady state.

```
plt.plot(demand.transpose())
plt.grid(True)
plt.xlabel('day')
plt.ylabel('probability of demand')
plt.gca().legend(('high','medium', 'low'))
plt.show()
```

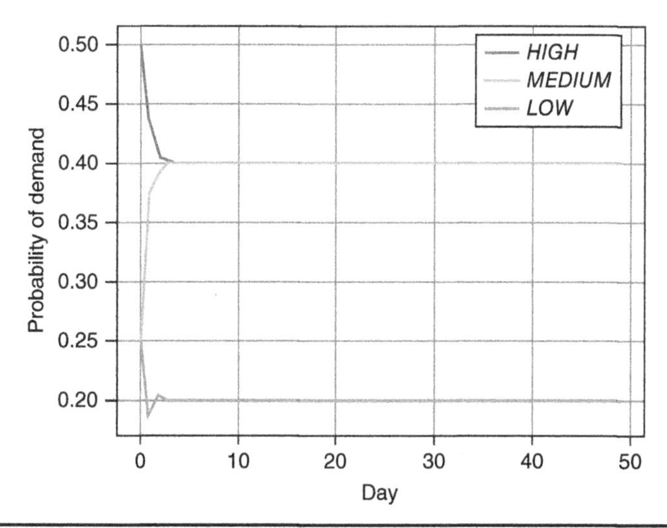

FIGURE 5.8 A transition diagram model for demand prediction using the Python code in this chapter.

References

Al-Anbuky, A., Bataineh, S., and Al-Aqtash, S., "Power Demand Prediction Using Fuzzy Logic," *Control Engineering Practice*, Vol. 3, No. 9, pp. 1291–1298, 1995.

Bakirtzis, A. G., et al., "Short Term Load Forecasting Using Fuzzy Neural Networks," *IEEE Transactions on Power Systems*, Vol. 10, No. 3, pp. 1518–1523, 1995.

Basu, S. N., "Short Term Localized Load Prediction," *Transactions on Power Systems*, Vol. 7, No. 1, pp. 389–397, 1992.

Bougaev, A., *A Methodology Using Support Vector Machines for Short-Term Load Forecasting*, MSNE Thesis, Purdue University, 2002.

Bougaev, A., Urmanov, A., Tsoukalas, L. H., and Gross, K., "Method of Key Vectors Extraction Using R-Cloud Classifiers," *Journal of New Mathematics and Natural Computation*, Vol. 3, No. 3, pp. 1–8, May 23, 2007.

Cox, E., "Fuzzy fundamentals," *IEEE Spectrum*, pp. 58–61, 1992.

Gross, G., and Galiana, F. D., "Short-Term Load Forecasting," *Proceedings of IEEE*, Vol. 75, No. 12, pp. 1558–1572, 1987.

Hobbs, B. F. et al., "Artificial Neural Networks for Short-Term Energy Forecasting: Accuracy and Economic Value," *Neurocomputing*, Vol. 23, pp. 71–84, 1998.

Mori, H., "Accumulative Effect of Discomfort Index for Fuzzy Short-Term Load Forecasting," *Engineering Intelligent Systems*, Vol. 7, No. 4, pp. 233–238, 1999.

Park, J. H. et al., "Composite Modeling for Adaptive Short-Term Load Forecasting," IEEE Transaction on Power Systems, Vol. 6, No. 2, pp. 450–457, 1991.

Peng, T. M., et al., "Advancement in the Application of Neural Networks for Short-Term Load Forecasting," *Transactions on Power Systems*, Vol. 7, No. 1, pp. 250–257, 1992.

Tsoukalas, L. H., "Neurofuzzy Approaches to Anticipation: A New Paradigm for Intelligent Systems," *IEEE Transactions on Systems, Man and Cybernetics*, Vol. 28, No. 4, pp. 573–582, 1998.

Tsoukalas, L. H., and Uhrig, R., *Fuzzy and Neural Approaches in Engineering*, John Wiley and Sons, New York, 1997.

Verona, F. B., and Ceraolo, M., "Use of Neural Networks for Customer Tariff Exploitation by Means of Short-Term Load Forecasting," *Neurocomputing*, Vol. 23, pp. 135–149, 1998.

Wang, X., Hatziargyriou, N., and Tsoukalas, L. H., "A New Methodology for Nodal Load Forecasting in Deregulated Power Systems," *IEEE Power Engineering Review*, Vol. 22, No. 5, pp. 48–51, 2002.

CHAPTER 6

Advanced Topics

Where we focus on fuzzy logic utilization in AI, real-world applications in power generation, forecasting, robotics, sensors, man-machine interface, and big data trends.

6.1 Fuzzy Logic and AI

For Alan Turing, the question "can machines think" was too meaningless to deserve discussion (Turing, 1950). The reason, he thought, was that we don't really understand what *thinking* is. Turing reformulated the question to what he called an "imitation game," known as the *Turing Test*, in which if a judge cannot distinguish between a computer and a human, both behind a wall and both furnishing typewritten answers to questions posed by the judge, then the judge may declare the computer as intelligent as the human. The Turing Test, a collaborative game between two humans and a machine, is what Wittgenstein would have called a *language game*. The setup can be read, at first, as a test for machine intelligence, although upon deeper reflection, may be seen as a practical definition of a type of intelligence relevant to innovation; where to anticipate the success of a new system may be analogous to benchmarking it against the best of peers by stripping away complexities and nuances not relevant to success. Innovation intelligence is facilitated by machines, in what we have come to call Artificial Intelligence or simply AI, and it consists of programming protocols enhancing the predictive process of humanity.

Nearly three decades after Turing, Zadeh proposed that if a computer is equipped with fuzzy logic, it could do better on a Turing test, simply by improving its programming through the use of human language and what he called *approximate reasoning* (Zadeh, 1981). The point of course is that approximate reasoning and AI can make it easier for programmers, we may think of them as "coaches," to improve on prediction, planning, control, monitoring, and decision-making.

Over the years, two major schools of thought have emerged in AI: the *logical* and the *statistical*. Statistical approaches aim at machine learning (ML) algorithms using data to create and refining models automatically. They include, but are not limited to, neural approaches, support vector machines, Bayesian and belief networks, and several game-theoretic methods. On the other hand, the logical school includes logic programming, fuzzy reasoning, and language-games that encode reasoning tasks with expert knowledge as well as strategy for resolving conflicts on the face of multiple choices or alternatives.

Fuzzy logic has been successfully applied to many AI areas, including knowledge-based systems, pattern recognition, human-machine interfaces, robotics, decision making, and control. If a process is too difficult to model precisely, a set of *if/then* rules may provide an easy and interpretable formulation. As it happens, modeling uncertainty (often called *epistemic uncertainty*) may be too high. Problems are often presenting us with too many variables and hence it is difficult to abstract away to a smaller set of variables, which would allow for the principal relations to be identified. A nuclear reactor, for example, can be modeled by 5000 variables or by five variables and any number of variables in between. In most instances, a smaller set of variables may result in a better model of the reactor. The issue here has to do with what is meant by "better" and in more than one ways this question devolves into issues of modeling uncertainty which in turn lead to fuzziness, in the sense that better means "better than other models" in enabling us to foresee.

In addition to modeling uncertainty, a problem may be fraught with significant data uncertainty (also called *aleatory uncertainty*) resulting in unreliable or faulty solutions. Such problems we encounter in financial, economic, or even medical applications. In these situations, fuzzy logic can be a successful bridge between language and artificial intelligence for actionable, corrective, and interpretable approaches. More recently, language processing has seen such notable successes with software that mimic human dialogues so well that the old Turing Test appears to be easily passed by machines doubling as human interlocutors.

AI enhances human capabilities to foresee and forecast the future through good feature detection and accurate analysis. In financial engineering, if some aspect of the future can be accurately anticipated (not precisely), one would know how, where, and when to trade, to invest, or whether to bet in one direction or another (Pantazopoulos, 1998). To make this kind of intelligence available calls for structured, but largely open, access to data as well as computers capable of facilitating domain-specific dialogues amongst traders. After all, the goal is to build capacities for strategic thinking, which is thinking that pertains to selecting from many different options, scenarios, or even goals for the future.

The breadth and diversity of fuzzy methods in AI are too numerous to capture in a brief survey. Our aim is to provide some illustration of how language as a tool for computing is more suited for advancing human-centered innovation. This includes ML, human-machine interfaces, robotics and automation, logistics, pattern recognition, image recognition, and cyber security.[1]

6.2 Fuzzy Control

Control is an area of fuzzy logic, immensely successful, and thus exercising a paradigmatic influence in many different domains of fuzzy logic applications. It is as if the language-game of control spills over to provide terms and phrases useful to language-games for monitoring, diagnostics, decision-making, and planning, to name but a few. To make things a bit more transparent we will describe here an example of fuzzy control for a small digital nuclear reactor, that is, the Purdue University Research Reactor One (PUR-1), the first nuclear reactor in the United States to be licensed with a completely digital Instrumentation and Controls (I&C) system. This work is described in detail by Oktavian et al. (2022).

An image of PUR-1 and a top-view schematic of its geometry can be seen in Fig. 6.1, where in (a) we have a photograph of the reactor from the top of the pool and in (b) we show the location of the control rods. The dynamic behavior of the reactor is described by the point kinetics equations (PKEs) (Lamarsh, 2001):

$$\frac{dP(t)}{dt} = \frac{(\rho(t)-\beta(t))}{\Lambda(t)} P(t) + \frac{1}{\Lambda_0} \sum_k \lambda_k \zeta_k(t) + \frac{1}{\Lambda(t)} S_d(t) \tag{6.1}$$

$$\frac{d\zeta_k(t)}{dt} \dot{\zeta}_k(t) = -\lambda_k \zeta_k(t) + \frac{\Lambda_0}{\Lambda(t)} \beta_k(t) P(t), \quad k = 1 \text{ to } 6 \tag{6.2}$$

Table 6.1 explains the meaning of the symbols used in Eq. (6.1), and Table 6.2 gives the values of the parameters used in Eq. (6.2). It should be noted that Eq. (6.2) is actually describing six equations, one for each k value. These six equations give the dynamic behavior of six different groups of delayed neutron precursors produced by nuclides born out of the fission process with a significant delay relative to the time of fission. The overall population of neutrons is principally responsible for the power of the reactor, the dynamics of which are described by Eq. (6.1). The overall population is directly related to the number of fissions per unit time by the chain reaction. The delayed neutrons can be thought of as the neutron supply at the margin, that is, a small number of neutrons in the order of 1% of the overall population, which because of their delayed appearance makes the reactor controllable via control rods.

[1]A detailed survey of early applications can be found in the English translation of the Japanese text *Applied Fuzzy Systems* (Terano, 1994). Additional early applications can also be found in Tsoukalas (1997).

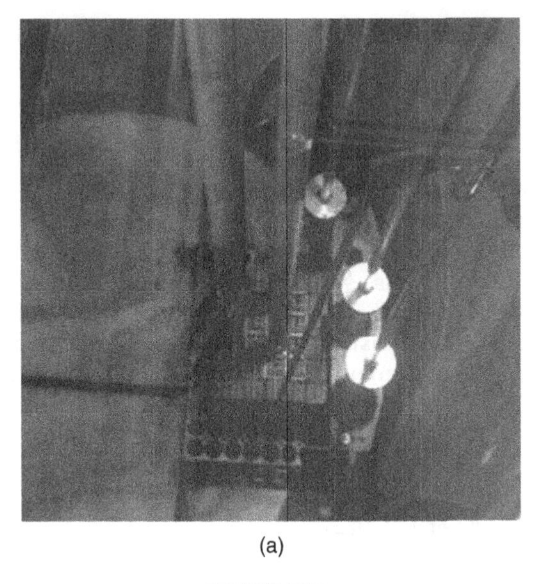

(a)

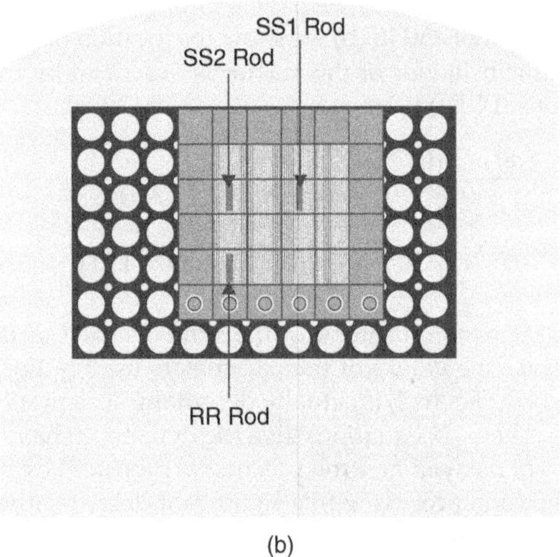

(b)

Figure 6.1 PUR-1 nuclear reactor (a) top view in the pool, (b) top view schematic of core geometry showing the locations of the control rods of the reactor.

For a small reactor of the type considered here, control rods are the main effectors of control. They inject or remove reactivity (that is, neutron multiplication through fission) of the reactor. Thus, the means of controlling the power level of the reactor, the time dependence of which is described by Eq. (6.1), is found in the control rods. The goal, of course, is to control the power level. Any planned raise or reduction of the power level is called a *power maneuver, startup*, and

Variable	Explanation
$P(t)$	Unnormalized power/flux level
Δt	Time-discretization time step
$\zeta_k(t)$	*k*th precursor density
β_k	*k*th delayed fraction
Λ	Neutron generation time
λ_k	*k*th group decay constant
S_d	Delayed source term

Table 6.1 Parameters Describing Reactor Dynamics

Group	λ	β (%)
1	0.0128	0.02584
2	0.0318	0.15200
3	0.119	0.13080
4	0.3181	0.30704
5	1.4027	0.11020
6	3.9286	0.02584

Table 6.2 Six-Delayed Group Parameter Values

shutdown being the basic power maneuvers, but additional maneuvers may be needed. All of them require steadfast attention to the physics of the fission process, certified operator credentials and expertise, high-level training, planning, and compliance with rigorous safeguards ensuring safety under all possible conditions.

The processes of fission, the so-called "chain reaction," run with its own physics clock, but reactor instruments and controllers run on time steps that, while in conformance with the physics of fission, are running on their own I&C clocks, which provide the time steps of interest to the controller (typically in the order of a second). At every time step, the necessary information is collected for the variables of primary interest, that is, neutron precursor concentrations $\zeta_k(t)$ of Eqs. (6.2) for six different groups and neutron flux $\phi_n(x)$. Neutron flux is a scalar quantity that measures the number of neutrons passing through an infinitesimally small area around any point inside the reactor in a small interval of time. It is a measure of the total numbers of neutrons available for fission, or *multiplication* as it is called, and it is given in units of neutrons (per cm^2 per second). Another quantity of interest is the reactor period, T, which is a measure of how fast these multiplications can grow if the reactor departs from equilibrium or steady state. The reactor period is the time required to change the reactor power by a factor of e (where e is the basis for natural logarithms, that is, $e = 2.71828$). If the reactor period is positive, the power of the reactor is increasing; if it is negative, it is decreasing. If power is increasing, the reactor period must be long enough in magnitude to prevent a dangerous excursion of reactor power. All reactors have automatic safety systems to immediately shut down the reactor if the period becomes too short. The reactivity of the reactor is given by

$$\rho = \frac{\Lambda}{T_t} + \sum_k \frac{\beta_k}{1 + \lambda_k T_t} \tag{6.3}$$

A cluster of fuzzy *if/then* rules, which in our example is called Fuzzy Rule-Based System (FRBS), is at the heart of the control architecture, as shown in Fig. 6.2. This fuzzy controller couples the fuzzy algorithm with Eqs. (6.1), (6.2), and (6.3) and uses Mamdani implication φ_c to model the rules. After rules are fired, it defuzzifies using *Center of Area* (COA) defuzzification (see 4.3).

The *LHS* fuzzy variables of the rules are *Power Error, Reactor Period* while *Reactivity* is the only *RHS* variable. The resultant contribution of the *RHS* values of rules having *Degree of Fulfillment* (*DOF*) other than trivial, when aggregated, provides a resultant membership function, which after defuzzification is turned into the control signal for the mechanism driving the control rods; in our case, we use only one control rod, whose position can be changed by inserting (insertion) or withdrawing it from the reactor core.

The fuzzy rule-based inference system is developed based on an expert understanding of the reactor. The complete set of *if/then* rules used to manipulate the output reactivity control signal is provided later in this chapter in Fig. 6.7, illustrated as

if *power error* is *EBN* **AND** *period* is *CP* *then* *reactivity insertion* is *RZ*
ELSE
if *power error* is *EBN* **AND** *period* is *LCP* *then* *reactivity insertion* is *RZ*
ELSE
. . .
if *power error* is *ESP* **AND** *period* is *PTI* *then* *reactivity insertion* is *RSN*

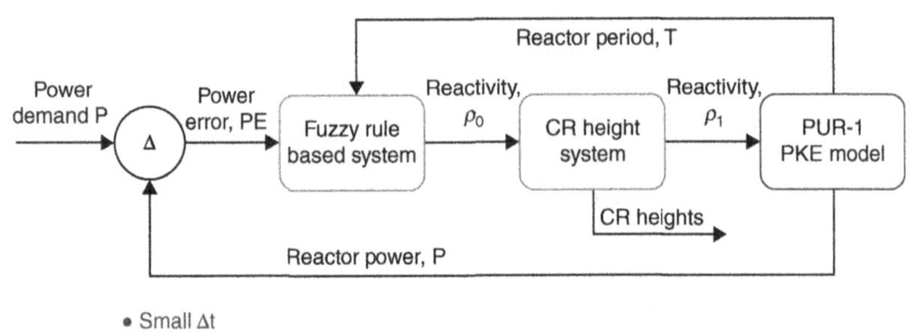

Figure 6.2 Fuzzy controller for nuclear reactor (Oktavian, 2022).

Inferencing is achieved through *max-min composition*. Since Mamdani implication is used, the aggregation operator $\varphi_\beta = \vee$, that is, *max* is used. *COA* defuzzification transforms the resultant membership function from the aggregation process into a single scalar, which is sent as an instruction to the actuator moving the control rod.

For the fuzzy *if/then* rules, two input variables and one output variable, a total of *three* fuzzy variables, are used. The fuzzy variables are as follows:

- Input 1: *power error* (Measured—Demand Power, in %)
- Input 2: *reactor period* (T, in *seconds*)
- Output: *reactivity insertion* (ρ, in $)

The fuzzy values for the error in reactor power are presented in Fig. 6.3. The fuzzy sets consist of five membership functions defined using trapezoidal and triangular plots. The zero error is defined as *EZ* to represent the ideal condition of the reactor when the measured power is quite close to the power demand. The *EZ* value ranges from –5% to 5% error with a peak in the 0% error. As many as five membership functions are used for the reactor period fuzzy sets, as shown in Fig. 6.4. Since the reactor period represents how fast the reactor power changes as a response to the reactivity insertion, the fuzzy sets are used to control this rate. For example, in the critical and little critical

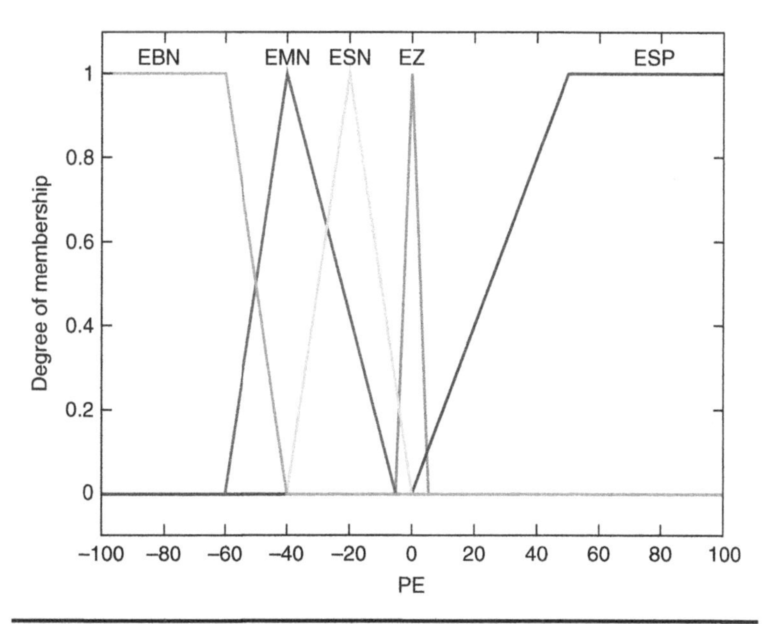

FIGURE 6.3 Values for the input variable *power error*: *EBN* (*error big negative*), *EMN* (*error medium negative*), *ESN* (*error small negative*), *EZ* (*error zero*), and *ESP* (*error small positive*).

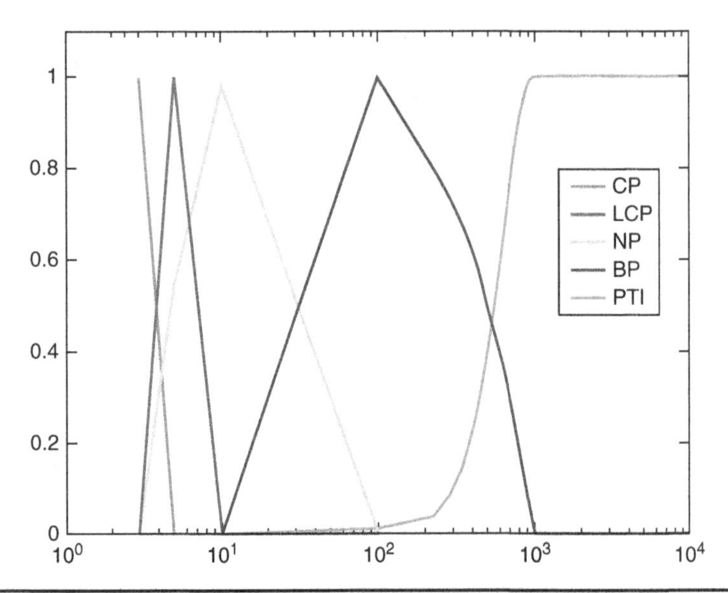

FIGURE 6.4 Values for the input variable *reactor period*: *CP* (*critical period*), *LCP* (*little critical period*), *NP* (*normal period*), *BP* (*big period*), and *PTI* (*period tends to infinity*).

periods, a positive reactivity insertion should be avoided at any cost. At this level, the reactor power can increase dramatically in just a few seconds.

For the *reactivity insertion RHS* variable (the output of the controller), six different membership functions are used for the fuzzy sets as shown in Fig. 6.5. The values range from −0.4$ to 0.4$ (reactivity is measured in "dollars" hence the $ sign is used here as a unit of reactivity) to avoid the reactor period being very small and unsafe for reactor operation (Fig. 6.6). The reactivity insertion impacts the reactor period directly. The 0.4$ limit is chosen for the reactivity insertion limit because the period in this value will be around 10 seconds, which is just a little bit above the critical period. Since there are five values for both *power error* and *period*, the fuzzy rules consist of 25 conditions (5×5). As shown in Fig. 6.7, the fuzzy rules include all the input fuzzy values defined. For safety reasons, there is no positive reactivity insertion in the critical period and little critical period. On the other hand, if the period is going to be infinitely large, a bigger reactivity insertion is possible.

The fuzzy algorithm is tabularly depicted in Fig. 6.7. It implements a control strategy similar to what is done during manual operations. Experimental data recorded from several startups, shutdowns, and power maneuvers were analyzed to determine the maximum power change rate utilized in actual operations along with the

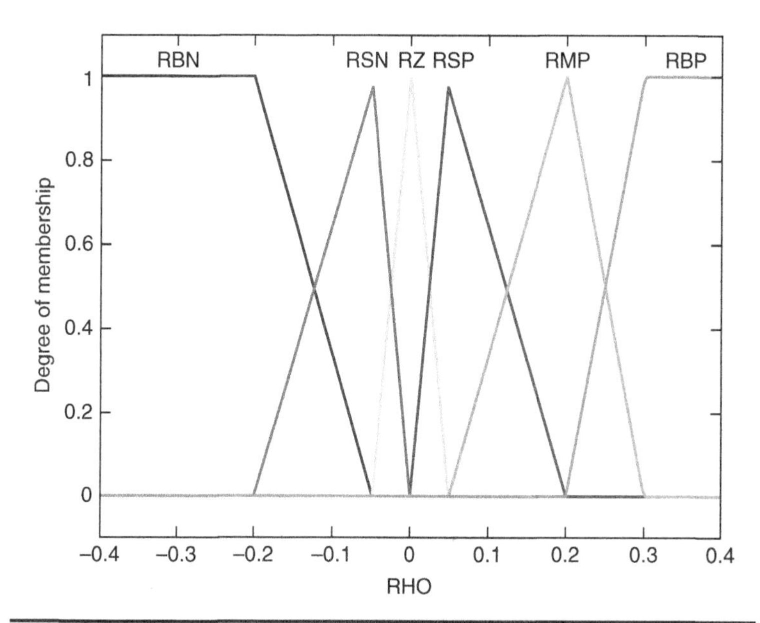

FIGURE 6.5 Values for the variable *reactivity insertion* (in \$ units of reactivity): *RBN* (*reactivity big negative*), *RSN* (*reactivity small negative*), *RZ* (*reactivity zero*), *RSP* (*reactivity small positive*), *RMP* (*reactivity medium positive*), and *RBP* (*reactivity big positive*).

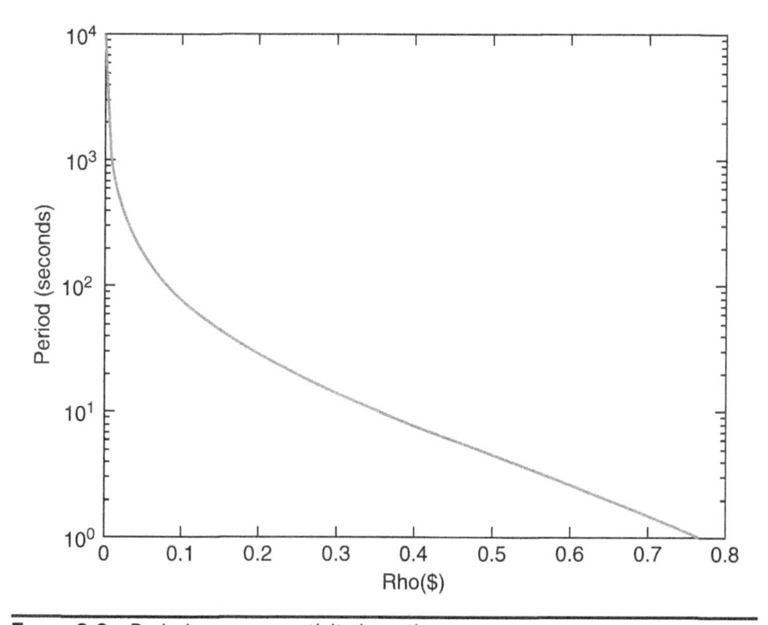

FIGURE 6.6 Period versus reactivity insertion.

		Period fuzzy sets				
		CP	LCP	NP	BP	PTI
Power error fuzzy sets	EBN	RZ	RZ	RSP	RMP	RBP
	EMN	RZ	RZ	RSP	RSP	RMP
	ESN	RZ	RZ	RZ	RSP	RSP
	EZ	RSN	RZ	RZ	RZ	RZ
	ESP	RBN	RSN	RSN	RSN	RSN

Figure 6.7 The *if/then* rules of the control algorithm in tabular representation with given input and output fuzzy values.

typical control rod (CR) movement procedures used to achieve startup and shutdown. All needed parameters were recorded by the reactor digital I&C systems and plotted to observe patterns.

An important limitation in operations is the conservative nature of changing the power level. During normal operations, a power change rate of 2%/s is rarely exceeded, which means going from zero to full power can take up to 15–20 minutes. Due to this, the maximum reactivity the algorithm can call for is limited to about 180 pcm, which is the amount of reactivity needed to initiate a roughly 2%/s power change rate.

It should be noted that only control rod RR (see Fig. 6.1 [b]) is used for the various power maneuvers. Control rods SS1 and SS2 are typically not used for power level changes; so all power changes are performed using the RR rod. These operating procedures can be seen in the PUR-1 experimental data for a simple power up from zero to near full power seen in Figs. 6.8 and 6.9.

To convert the reactivity requested by the fuzzy controller to CR positions or heights, it is necessary to know the rod worth curves for all three rods. All three CRs for PUR-1 have a maximum movement of 62.5 cm. From experimentation, the integral rod worth curves for SS1 and SS2 are known. However, only the integral rod worth of the RR rod is unknown (Townsend, 2016).

Figure 6.10 shows the flow chart algorithm used for the FRBS automatic CR system. The operational test cases include the startup operation, power maneuver, and shutdown. These three tests are important

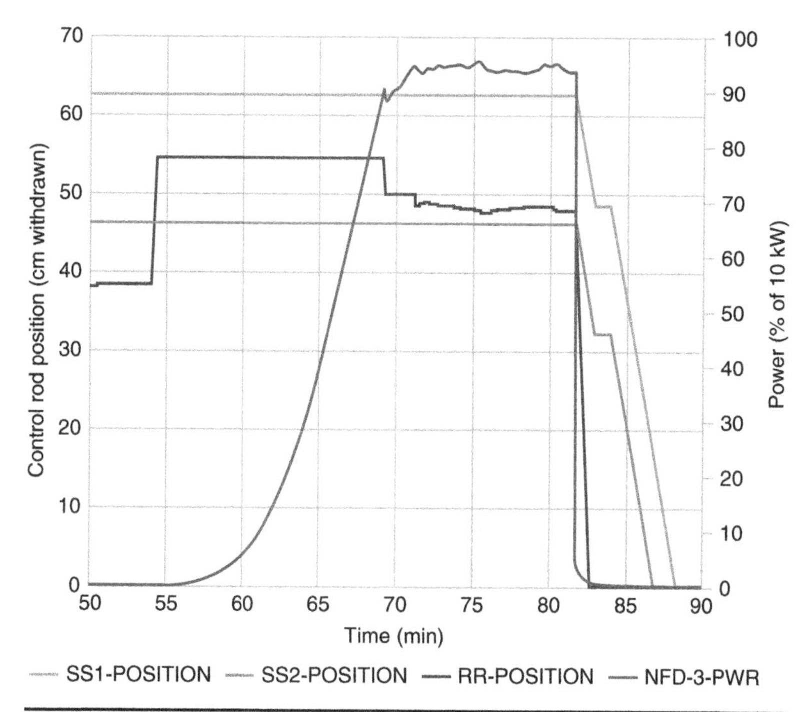

FIGURE 6.8 Experimental CR positions versus power.

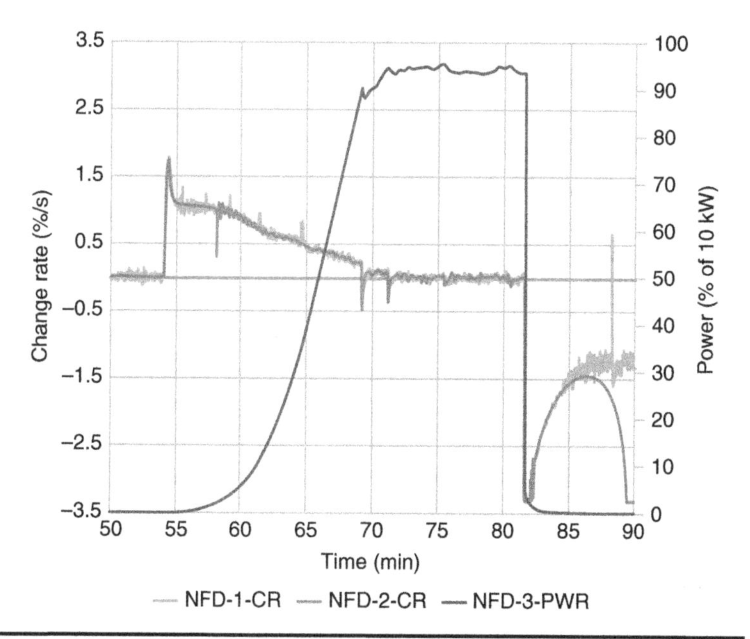

FIGURE 6.9 Experimental power change rate versus power.

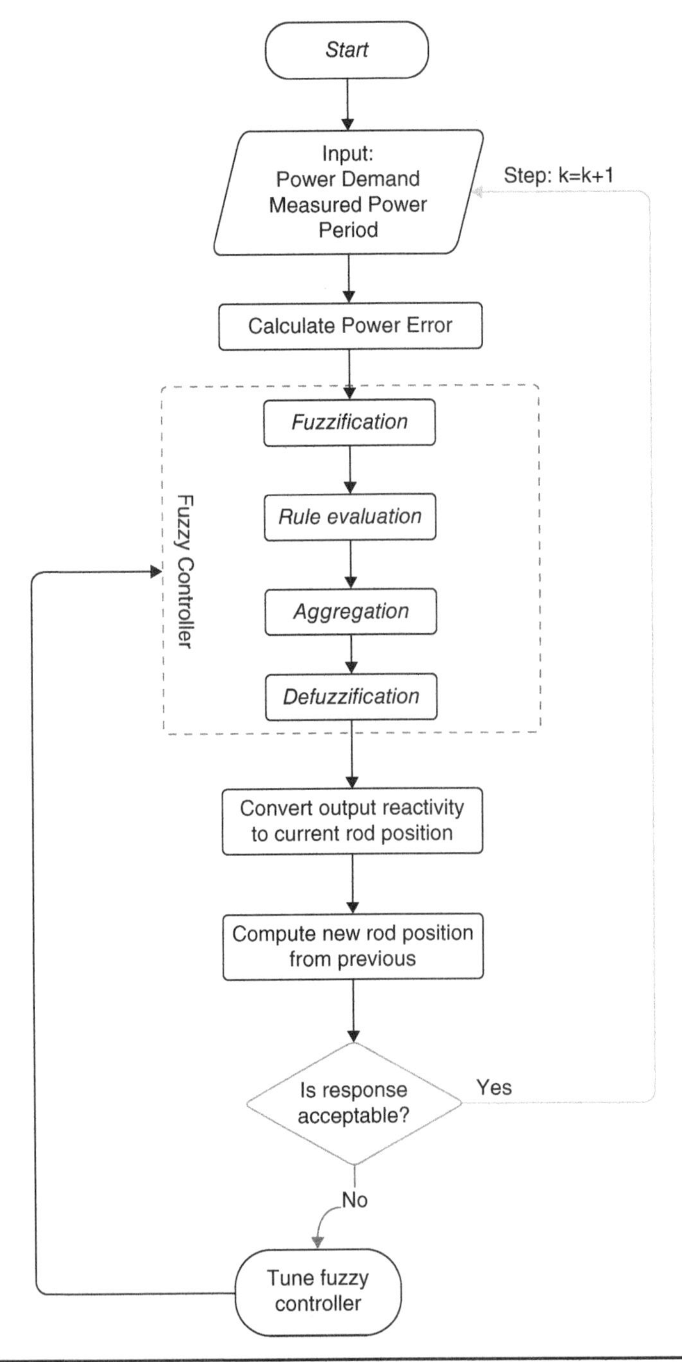

FIGURE 6.10 Automatic control rod (CR) system algorithm.

to the reactor operation and cover all the necessary parts of the power controller, to increase, decrease, and maintain the reactor power.

Reactor Startup

The plots of power and reactivity insertion evolution for each minute of the startup process are presented in Fig. 6.11. This case illustrates the reactor startup in PUR-1 going from zero power to full reactor power.

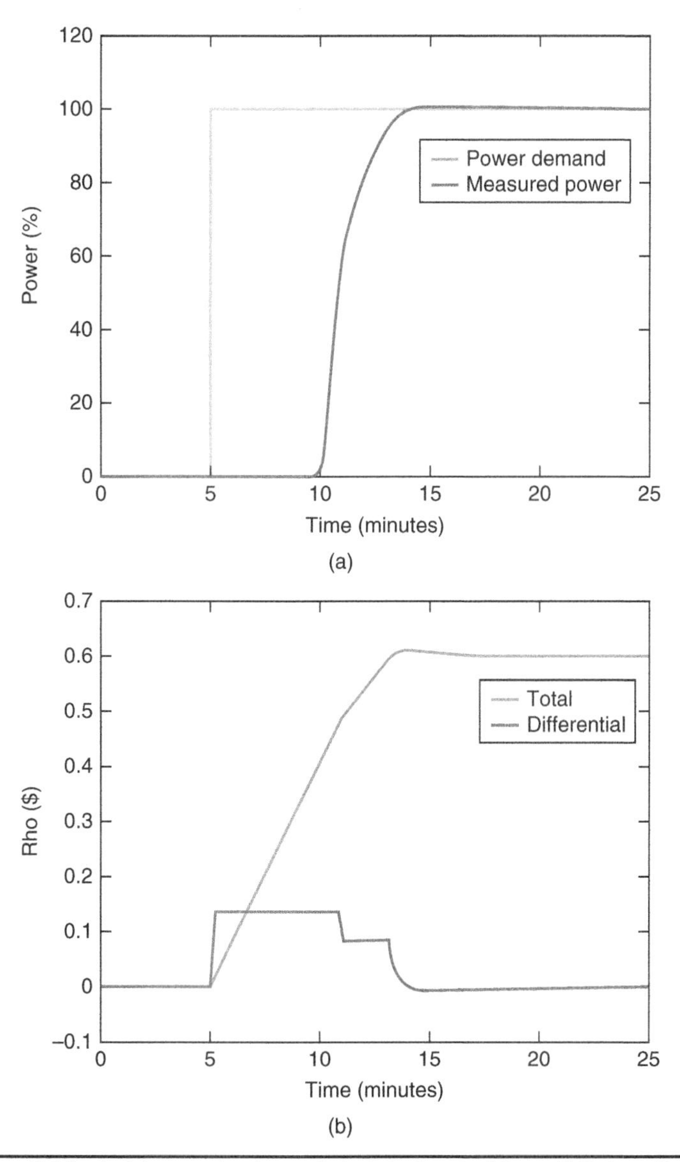

FIGURE 6.11 In (a) power is shown during startup; in (b) reactivity insertion during start up.

The demanded power was set to 100% at 5-minute time intervals. Currently, when this process is conducted, the PUR-1 reactor operator will need to manually adjust the CR position to give a positive reactivity insertion. The fuzzy controller is adjusting the measured power to follow the power demand automatically as shown in Fig. 6.11(a).

In the CR system, only the RR rod is moved to initiate changes in reactivity. If the RR rod reaches its maximum amount of movement and has not yet achieved the desired reactivity, the SS1 rod moves a preset amount to allow the RR rod to reset, allowing for additional room for changes in reactivity. In Fig. 6.11(b), the reactivity insertion for each time step is given during reactor start up. Initially, a positive reactivity insertion is given to increase the reactor power. After reaching the demanded power, the reactivity insertion is decreased eventually back to 0.0$ to further stabilize the constant power. The CR height evolution is presented in Fig. 6.8. It can be observed that this follows a similar shape with the reactivity insertion and the power experimental plot shown in Fig. 6.11(b). On a side note, only the RR (regulating rod) moved in this case, resulting in sufficiently slow, but safe power changes. To provide positive reactivity insertion, the CR is raised from 30 cm to approximately 42 cm of height. After the power level reached the demanded power, the CR is returned gradually to its initial position.

Power Maneuver

Unlike startup and shutdown, a power maneuver from 20% to 80% and then back to 20% presents its own unique challenges. We use them to assess the capabilities of fuzzy controllers to follow power demand changes dynamically. To demonstrate the flexibility of the fuzzy controller to respond to the power maneuver, a test is conducted as presented in Fig. 6.12. In this instance, after 5 minutes of reactor operation at low power (20%), the power is demanded to increase to a significantly higher level (80%). After around 20 minutes, the power is set to decrease again to 20% power. In less than 10 minutes, the reactor can achieve the power demand for increasing or decreasing power. In the plot, we can also see that there is an overshoot and undershoot in the power level. However, the fuzzy controller can fix this quickly to maintain the reactor power to the demanded or reference power level.

Shutdown

For the final test, a demonstration for reactor shutdown is conducted as shown in Fig. 6.12. It can be observed that the reactor can reach zero power level in around 15 minutes from the full power. With the fuzzy controller, there is no need to manually adjust the power level using CRs. Instead, the fuzzy controller will automatically and slowly reduce the power level until zero power is achieved and the reactor can be shut down safely.

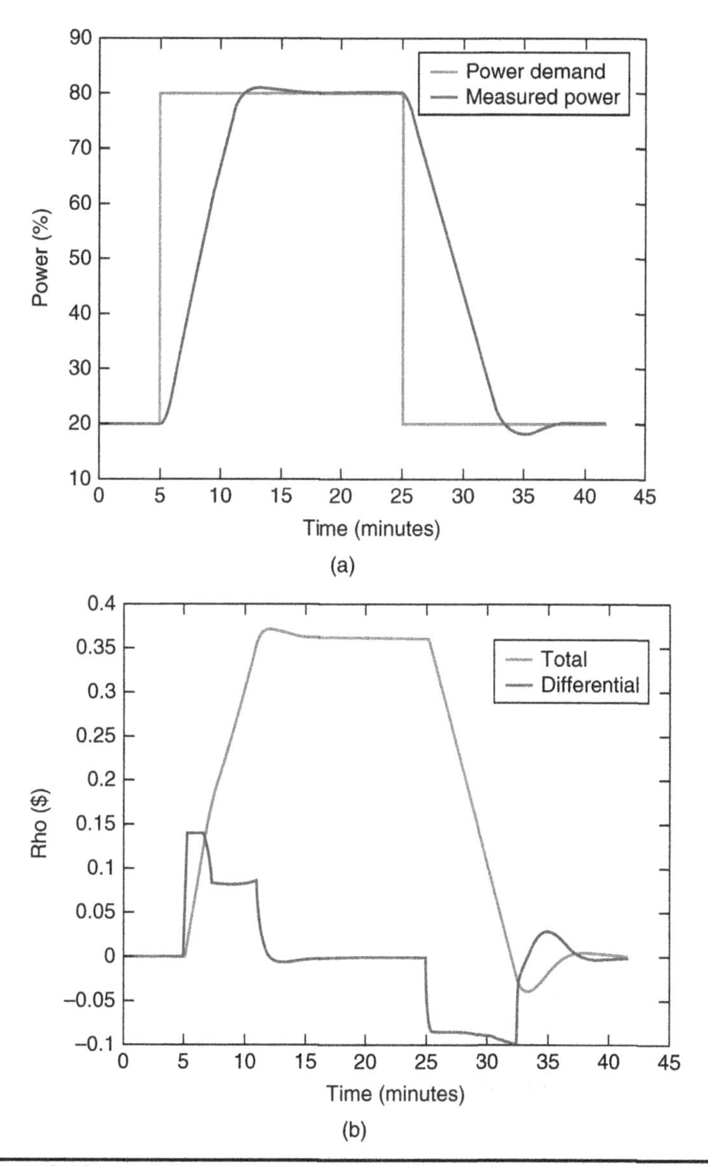

FIGURE 6.12 Power evolution is shown in (a); reactivity insertion is shown in (b) in the case of power maneuver from 20% to 80%, then back to 20.

Autonomously performed power maneuvers are *sine qua non* for reactor digital twins (DTs), as is also some capacity for learning. This can be extremely useful for clusters of micro reactors or small modular reactors, which need to be monitored remotely. The entire reactor behavior can be learned by its DT and in this way monitoring the reactor in its entire life cycle including its fuel elements through the analysis of big data, which would effectively address all concerns related to nuclear safety, security, and safeguards.

The fuzzy controller can be used to control the power output in a new digital I&C reactor to maintain a fail-safe condition during the automatic operation of the reactor and prevent possible operator errors. A simple set of rules, based on expert knowledge of reactor operations, can achieve power control in place of an operator. The fuzzy controller guaranteed smooth transitions during possible transients with inherent stability encountered during operation of the research reactor under study. This system should be able to eventually replace the manual control system after the fuzzy parameters have been tuned based on the heuristic membership functions in the inference system and form the core of nuclear reactor DT.

6.3 Neuro-Fuzzy Control

This section illustrates the capabilities of coupling an adaptive neural network to the fuzzy power controller developed for PUR-1 described in the previous section of this chapter. The purpose of this work done by Appiah et al. (2022) is to assess the essential benefits of coupling intelligent hybrid systems as real-time monitoring control mechanism for reactor fuel lifetime and power stability through the use of big data to address nuclear safety concerns. The adaptive network fuzzy inference system (ANFIS) is a feedforward network that combines both neural networks and fuzzy logic to show the relationship between inputs and outputs. The advantages are that it utilizes neural network learning algorithms to tune parameters, is good for nonlinear fitting and prediction with large data, and compares well with other artificial neural network methods. The ANFIS system was also implemented as a MATLAB code with two inputs, namely power error (PE) and period (T), and one output prediction for power or reactivity (ρ). The learning algorithm used for the ANFIS model is the hybrid-learning algorithm, which is the combination of gradient descent and least-squares methods using the forward and backward pass. The neural network learns and trains for parameter values that can sufficiently fit the training data. Each epoch of the hybrid learning algorithm consisted of a forward pass and a backward pass. The layers in Fig. 6.13 are explained as:

- **Layer 1**: This is the input layer. Each crisp input is assigned to a node and moves to the next layer.
- **Layer 2**: Also known as the fuzzification layer, where all nodes are adaptive. Outputs of this node are the fuzzy membership grades of inputs.
- **Layer 3**: The rule layer neuron receives inputs from the fuzzification neurons and computes the firing strength of the rule it represents using the AND operator.

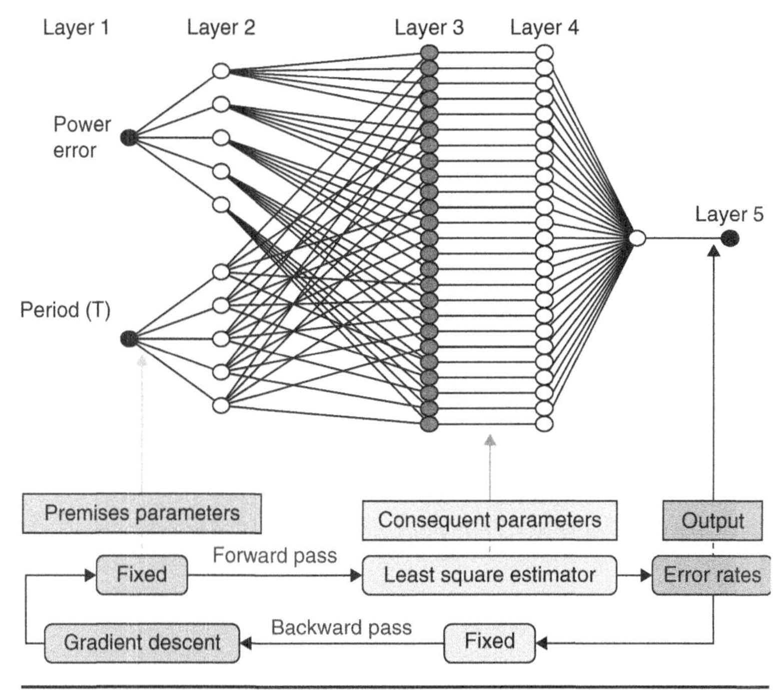

FIGURE 6.13 Developed ANFIS network structure.

- **Layer 4**: The normalization layer computes the normalized firing strength of each input signal from the rule layer.
- **Layer 5:** This performs the summation of incoming signals to get the final output. The weighted firing strengths and polynomial outputs (linear combination of inputs and outputs) associated with each rule *n*, respectively.

ANFIS Optimization with Genetic and Particle Swarm Algorithm

The adaptive neuro fuzzy predictive controller feedback loop is depicted in Fig. 6.15. A numerical optimization technique is used to determine the power control value $u(k)$ within the control time step. The minimization criterion Γ is given by:

$$\Gamma = \left\{ \begin{array}{l} \sum_{i=M_1}^{M_2} [r(k+1) - \hat{y}(k+1)]^2 \\ + w \cdot \sum_{i=1}^{M_u} [\Delta u(k+i-1)]^2 \end{array} \right\} \longrightarrow min$$

$$u(k) = u(k-1) + \Delta u(k) \tag{6.4}$$

The reference signal r in this case is the power demand, $\hat{y}$ is the predicted reactor power, and Δu is the control value change output from the fuzzy controller. Control performance depends on M_1 and M_2, which are the limits of the output prediction, W the weight constant, and M_u, which is the limit of the control value. The genetic and particle swarm algorithms are used to determine the optimal values of the predictive controller.

The genetic algorithm (GA) is a probabilistic optimization search approach for finding an optimal solution. An overview of the implemented algorithm in ANFIS is given in Fig. 6.14. The particle swarm optimization (PSO) also uses a similar search criterion like the GA. However, it uses the crossover selection and mutation to iteratively update the local and global values to obtain an optimal solution. The implemented algorithm in ANFIS is described in Fig. 6.15.

Simulation Parameters

The adaptive predictive controller proposed in this work is used to predict the startup output power of PUR-1 from reactor startup till steady state without overshoots and oscillations with respective simulation parameters given in Table 6.3. The simulation dataset was split into training and testing sets in the percentage of 80% and 20%, respectively.

Table 6.4 shows the favorable comparison between the FLC output and the ANFIS/GA, ANFIS PSO, ANFIS baseline model. The predicted results from the ANFIS PSO & ANFIS GA were very close with better

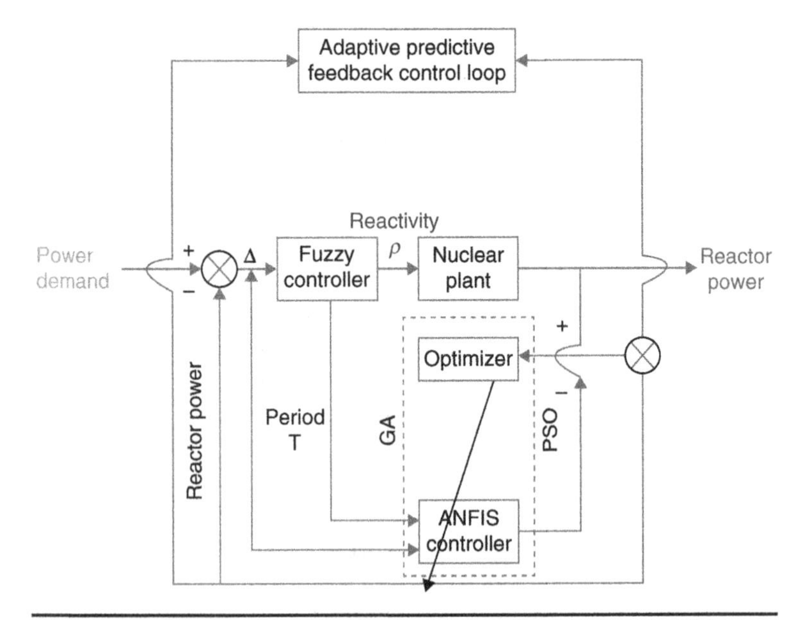

Figure 6.14 Developed ANFIS network structure (Appiah et al., 2022).

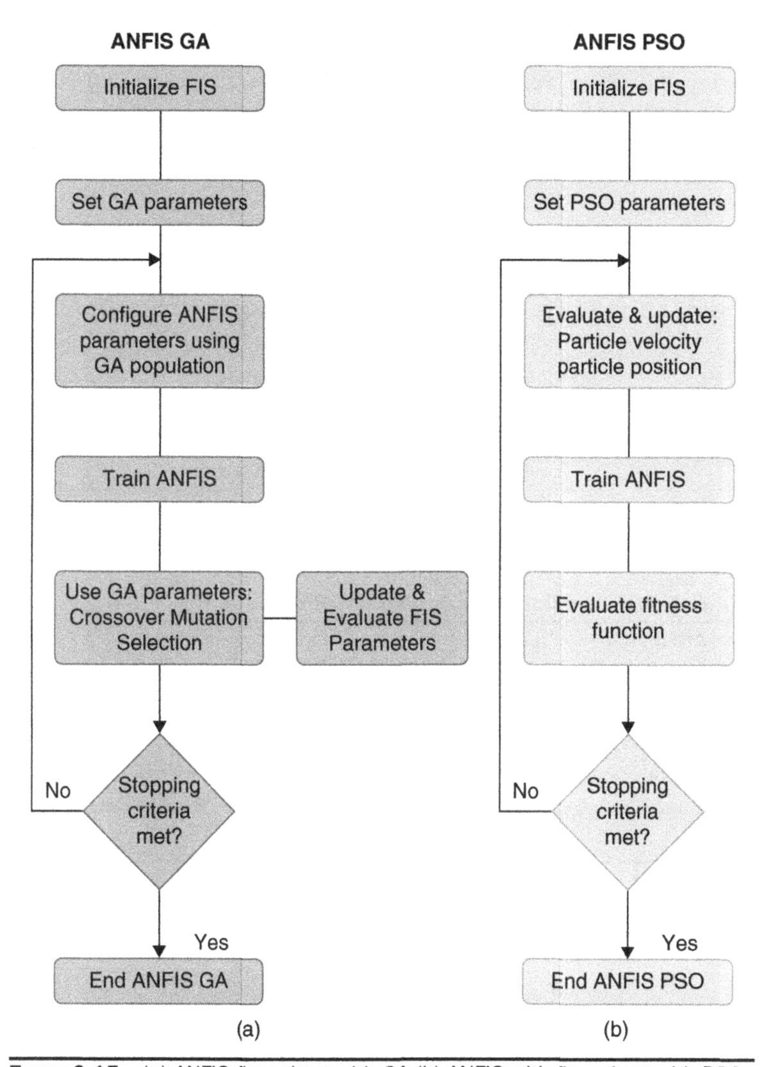

FIGURE 6.15 (a) ANFIS flow chart with GA (b) ANFIS with flow chart with PSO.

comparison to the ANFIS baseline controller output. The surface plot depicting the relationship between input and output parameters after training is shown in Fig. 6.16.

In Fig. 6.17, the predicted fuzzy power profile for PUR-1 reactor startup goes from zero to full demanded power (100%) in 5 seconds time stamp. Reactivity is inserted to increase the reactor power to the setpoint, and later reduced to further stabilize the power. The fuzzy controller system shows the feasibility of adjusting reactor power automatically to follow the power demand whilst the ANFIS controller initiates the predicted output power profile in the coupled intelligent system.

ANFIS Parameters		GA Parameters		PSO Parameters	
Number of nodes	75	Population size	100	Population size	100
Number of nonlinear parameters	40	Max. no. iteration	300	Max. No. iteration	300
Total no. of parameters	115	Crossover percentage	0.7	Inertia weight	1.0
Training data pairs	2400	Mutation percentage	0.5	Damping ratio	0.99
Testing data pairs	600	Pressure selection	8	Learning coefficient	1.0
Epochs	200	Gamma	0.2	Leaning coefficient	2.0

Table 6.3 ANFIS Optimization Simulation Parameters

Network	Output	Data	RMSE	R^2
ANFIS/GA	Power prediction	Train test	0.1620 0.5791	0.9015 0.8878
ANFIS/PSO		Train test	0.1530 0.0050	0.9925 0.9896
ANFIS		Train test	0.5592 0.6345	0.8683 0.8063

Table 6.4 Optimized ANFIS Controller Prediction Comparison

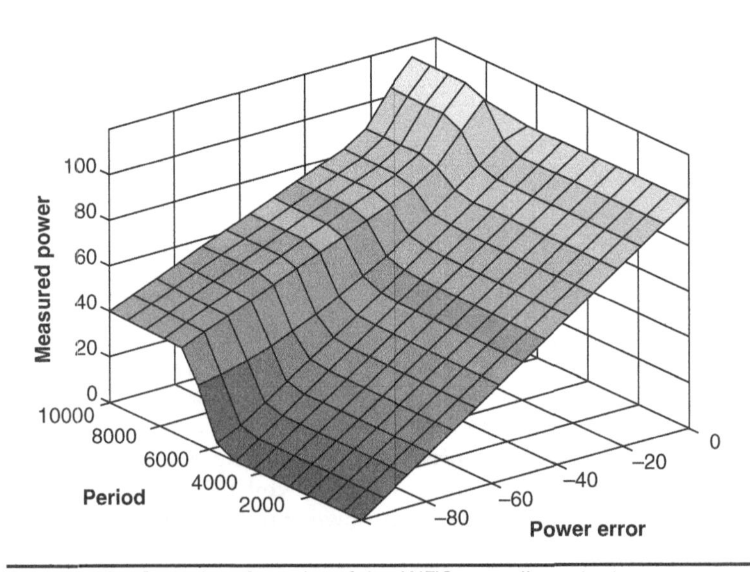

Figure 6.16 Control surface plot of the ANFIS controller output.

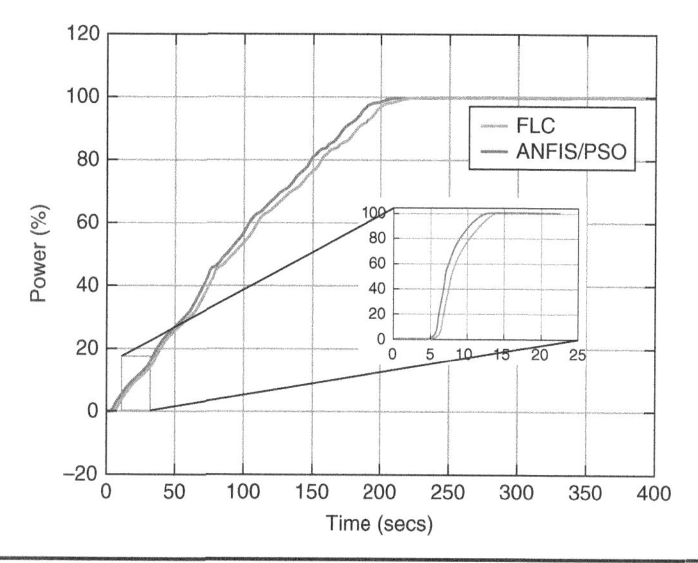

FIGURE 6.17 ANFIS/PSO prediction.

6.4 Digital Twins and Transfer Learning

The concept of a DT refers to computer programs that mimic a physical system, such as a nuclear reactor, with sufficient fidelity, reliability, and accuracy that the computer programs can stand for their physical sibling. A DT could *de facto* pass a Turing test, and thus be used as a general source of intelligence pertaining to its twin system and by that is meant intelligence that pertains to tasks and maneuvers including, but not limited to, planning, startup maneuvers, exercises, maintenance routines, monitoring, and analysis of performance.

In a new generation of nuclear reactors, such as micro reactors, producing up to 5 MW of power, or small modular reactors (SMRs) producing up to 300 MW of power, DTs are expected to play a major role. These reactors may be built in clusters which can be centrally monitored from afar with their DTs being indispensable doubles for all kinds of monitoring, diagnostics, control, planning and maintenance, and operations activities. The DTs could be customized to their physical twins via ML so that aging effects as well as various idiosyncratic characteristics that emerge and uniquely identify such systems in the life cycle can be absorbed by their digital siblings. SMRs may provide power outside the power grid, or as support to highly interconnected grids, effectively acting like uninterrupted power supplies (UPS) at grid levels in the sense that to avoid cascading failures, the grid can devolve into microgrids each supported by local micros or SMRs in a process known as islanding.

DTs are the digital representation of a physical asset or system that relies on real-time and history data for inferring complete reactor states, finding available control actions, predicting future transients, and identifying the most preferred actions. The concept of DTs has been implemented in the disciplines of health, meteorology, manufacturing, education, cities, transportation, and energy sector. More recently, DTs were implemented for process monitoring, and specifically for nuclear reactor monitoring through physics-informed neural networks (Prantikos, 2022). In addition to ML-based DTs, there has been development of DTs using fuzzy logic.

A neuro-fuzzy DT was developed for a high temperature generator (HTG) (Salazar, 2022). More specifically, a neuro-fuzzy approach is used to describe HTG, internal lithium-bromide temperature, and water outlet temperature. Then, using real data from 8 days of operation, two ANFISs are trained. The developed model was verified using 2 days of real data, and the findings were promising. The acquired DT's low validation errors imply a reasonable performance in capturing the dynamics of the HTG with adaptive capabilities, given that the learning step can be updated every day.

Digital Twin for Nuclear Reactor Monitoring

Advances in nuclear reactor performance efficiency can be accomplished using state-of-the-art monitoring capabilities. A concept of a DT has been proposed recently for process monitoring, including nuclear reactor monitoring (Prantikos, 2022). The DT consists of a computational model that tracks the history, and continuously adjusts the model to detect anomalies, such as degradation and insipient signs of failure of components, materials, and sensors. Proposed approaches for implementation of a DT involve either a physics-based differential equations model, or a data-driven ML model. The challenges of using model-based DT for nuclear reactor monitoring consist of accounting for a priori unknown loss terms in the complex experimental system to achieve sufficiently close agreement between the model and observations. Data-driven ML model captures the information about the reactor system from the experimental data used for model training. However, unlike the model-based DT which contains equations describing time evolution of the system, ML-based DT has limited extrapolation capability and, in principle, requires arbitrarily large amount of training data. Indeed, ML models follow an empirical, data-driven approach in making predictions based on large collections of historical data in order to achieve high performance. Although ML models are computationally fast in making predictions and robust with respect to noisy data, they are frequently difficult to both interpret and to develop from data. Furthermore, the data commonly need to be accompanied by labels that are not easily available. In addition, while ML models can be versatile to a varying extent and resolution of input data, their usual requirement of large volumes of training data hinders model development. On the other hand, physics-based models can alleviate the constraint of the big data, as the DT training no longer relies solely on the behavior of input-output examples. Moreover, DTs may have to be constructed for the system where there is scarce data availability. There exist methods such as few-shot learning, which offer the potential of learning from small datasets. However, in few-shot learning, the lack of information content in short temporal space is usually compensated with another dimension of the data (e.g., number of sensors). This imposed additional requirements on data collection.

Recently a hybrid approach has emerged consisting of physics informed neural networks (PINN) designed to solve differential equations. The PINNs address the need for integrating governing physical equations of the process into ML models, which establishes theoretical constraints and biases to supplement measurement data. The integration of governing physical equations into ML models provides solution to several limitations of purely data-driven ML models. First, in the case of data scarcity, most ML approaches are unable to effectively work, because there is a minimal required data volume to train the model. In the case of PINNs, the model can be trained without big data availability. Second, in the case of big data availability, ML approaches face severe challenges to extract interpretable knowledge. Furthermore, using purely data-driven models can lead to overfitting of the observations. As a result, this may introduce physically inconsistent predictions due to extrapolation or biases, and eventually result to low predictability. PINNs performance is not directly related to the volume of data, rather than the physics underlying the behavior of the system.

The governing equations for modeling a nuclear reactor are Boltzmann neutron transport and Bateman partial differential equations (PDEs). Numerical solution of these PDEs for a typical reactor geometry requires extensive computational resources. A reduced order model consisting of a system of point kinetics equations (PKEs) has been developed for the case when the spatial dependence of the neutron flux can be ignored, which is typically valid for small reactors. In this paper, the application of PINNs to numerical solution of PKEs which model the Purdue University Reactor Number One (PUR-1) small research reactor was investigated. Figure 6.18 shows a schematic of PUR-1. A PINN solution of PKEs using experimental parameters of PUR-1, such as values of the reactivity schedule and neutron source was developed. The PKE case investigated in this paper involves time-dependent stiff nonlinear ODEs, where the range of values is approximately eight orders of magnitude, during time interval of several hundred seconds. Results of this paper demonstrate strong agreement between PINN solution and numerical solution of PKEs

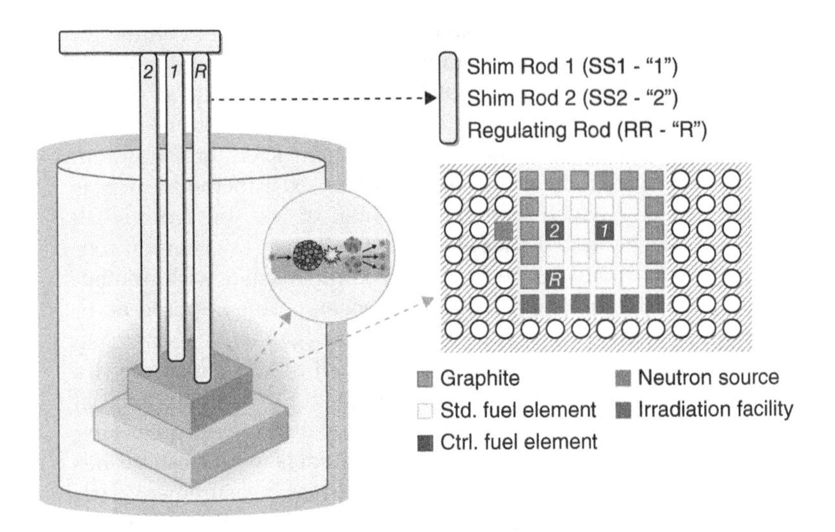

FIGURE 6.18 Schematics drawing of PUR-1. The inset panel 3 shows relative locations of the fuel elements and control rods.

using finite difference solver. Figure 6.19 shows PINN solution along with the numerical solution for the neutron density concentration, which is only one term of the PKEs.

A new approach for developing a nuclear reactor DT is based on PINN, which uses ML methods to solve governing differential equations

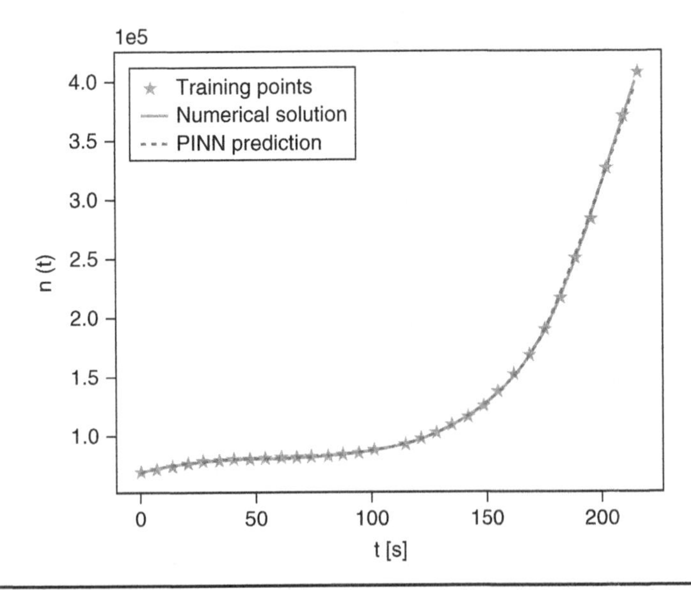

FIGURE 6.19 Solution of neutron density concentration $n(t)$ in the time interval $t \in [0, 217]$s along with PINN prediction.

(Prantikos, 2022). Using PINNs is a hybrid approach that provides alternatives to a purely physics model based-DT, which has challenges in modeling complex experimental systems, and a purely data driven ML-based DT, which relies on training data and has challenges with extrapolation. A PINN solves Eqs. (6.1) and (6.2) without the need for time-consuming construction of elaborate grids, offering low computational time. Hence, it can be applied more efficiently to solution in irregular and high-dimensional domains. Fast computations with PINN are particularly beneficial for real-time reactor operation monitoring. Given a reactivity schedule and initial values, PINNs execute rapidly on available desktops without the need for high-performance computing hardware (Prantikos, 2022).

6.5 Cybersecurity

Digitization of instrumentation and control (I&C) systems has brought many advantages including, but not limited to, measurement precision, reliability, faster processing times and reduction of cabling. However, digitation presents us with some drawbacks such as development or validation and verification (V&V) costs, risks of failure, and most importantly vulnerability against attacks. There is hence an immense need for methods toward ensuring the safety and security of these systems. All the characteristics mentioned above imply the category of cyber physical systems (CPSs), which combine features from both the computational, as well as the physical world. The accuracy of computing meets with the uncertainty of physical sensor inputs. The popularity and increased use of CPSs can thus enlarge cyber-attack risks pertaining to a system's integrity and security. Nuclear reactors can also be included in the category of CPSs. The process of digitization of several components of their I&C systems make the need for fortification in order to detect and mitigate attacks even more imminent. A work from a few years ago explored the establishment of cybersecurity in the Purdue reactor (PUR-1), which is the first all-digital nuclear reactor in the United States (Pantopoulou, 2021).

PUR-1 was physically modeled by using the PKEs. These equations describe the change of various reactors parameters with time. These equations are given in Eqs. (6.1) and (6.2), but a slightly different version used for cyberphysical security is shown below.

$$\begin{aligned} \frac{dn(t)}{dt} &= \frac{\rho(t)-\beta}{\Lambda}n(t)+\lambda c(t) \\ \frac{dc(t)}{dt} &= \frac{\beta}{\Lambda}n(t)-\lambda c(t) \end{aligned} \tag{6.5}$$

where n is the neutron density, c is the delayed neutron precursor density, ρ is the reactivity, β is the delayed neutron fraction, Λ is the mean neutron lifetime in the reactor core, and λ is the mean neutron precursor lifetime. Knowing the reactor state at each time point from these equations, one can easily calculate the power P of a reactor with the following formula:

$$P = V_r V_{\text{fuel}} n \bar{v} \Sigma_f E_f \tag{6.6}$$

where, V_r is the volume of the reactor, V_{fuel} is the percent amount of fuel volume (compared to the coolant volume), n is the neutron density, $\bar{v}$ represents the mean neutron velocity in the reactor, Σ_f is the macroscopic fission cross section, and E_f is the energy released per fission event. A proportional-integral-derivative (PID) controller was used to control the output power of the PUR-1.

The main approach to achieve cybersecurity of PUR-1 was based on an algorithm developed by the Office of Naval Research (ONR), called RHIMES (Mertoguno, 2019). The version of this algorithm studied for PUR-1 uses three (redundant) identical controllers and can maintain system operation even when threats are present. The two first controllers are used to detect a possible failure due to a malicious input. In case of a failure, the input that caused this to happen gets deleted and never reaches the third controller, so that the system can operate normally despite the present threat.

A graphical user interface (GUI) was furthermore implemented for visualization purposes. The results showed that PUR-1 was able to overcome malicious inputs regarding scram operation or regarding the positions of rods when the protection architecture was activated.

6.6 Transfer Learning for Trustworthy and Explainable AI

Transfer learning involves reuse of AI modules to solve similar, albeit different, problems; or, reuse for new and significantly different problems (Weiss, 2016). Reuse requires adaptation (statistical AI) and often strategy (logical AI). In fuzzy logic, such modules are typically, but not exclusively, collections of *if/then* rules. In statistical AI, they are neural networks, Bayesian systems, support vector machines or other types of modules where a large number of parameters are adjusted through machine learning (ML). As illustrated by the *Desert Island Thought Experiment* of Chap. 1 (p. 3), language is constructed in groups, where an individual through social interactions builds meaning and syntax in a process evolving against a *social-linguistic background*. The result is language-games, where new games reuse elements of older games. For instance, the language-game for potable water "spills over" to language-games for eatable fruits, safe shelters, training, or even entertainment. The idea of *reuse* (*transfer learning*) is actually embedded in the concept of *language-game*.

DTs are faithful mirrors of physical systems that ought to reuse modules or language-games as they grow and adapt to dynamic or structural changes. Transfer learning is a powerful approach for DT modules to acquire new knowledge but also, and most importantly, to address issues of *trustworthiness* and *explainability* which may limit their use in critical applications such as nuclear systems. How can this be done? The short answer is just like pupils learning different subject areas as they advance through school grades. Pupils pass exams to acquire certificates of performance. Similarly, AI modules can teach and be taught and take "exams" to acquire certificates of competence. We can think of this process as building on language-games acquired in a grade-like fashion where knowledge transfers from *simple to complex*, like it happens with pupils advancing through various grades (we call it *vertical transfer learning*) and particular subject areas of knowledge such as math, physics, and chemistry (we call it *horizontal transfer learning*). For instance, a student in the subject area of mathematics needs to master algebra before taking calculus or differential equations. At each level, the student engages in a language-game with a teacher. For instance, Calculus I requires Algebra II and a teacher attests to possession of required skills and competences through examinations. The teacher logic can be articulated in a set R^N of N fuzzy *if/then* rules. For example,

R^N:	*if equation_solving is A*	*then pre_calc is AVERAGE,*	*ELSE*
	if trig_identities is (A OR B)	*then pre_calc is POSITIVE_AVERAGE,*	*ELSE*
	… … … …		
	if conic_sections is (A OR B)	*then pre_calc is EXCELLENT*	

where, *equation_solving* refers to an *exam* on first- and second-order algebraic equations, *A* refers to the exam grade; *pre_calc* is a variable that signifies the readiness for calculus; *trig_identities* refers to testing for trigonometric identities; and *conic_sections* for circles, parabolas, ellipse, and hyperbolas.

DT modules can learn and teach each other in school-like fashion with protocols that result in quantifiably trustworthy and interpretable knowledge. Although many AI/ML algorithms used in DTs perform well for monitoring, maintenance, and scheduling tasks, the need to retrain them from scratch every time the system state changes significantly, makes it difficult, if not impossible, to deploy these algorithms in critical control or protection systems (including those for cybersecurity) under all possible conditions. Transfer learning is crucial for these modules to learn from each other and achieve a *"ratcheting effect,"* that is, preserving and correcting previously acquired knowledge while learning new things. For AI to be actionable, it must be trustworthy; and for trustworthiness to be quantified, there has to be *transfer learning*.

Consider, for example, the PUR-1 DT of Sec. 6.4 which mirrors a fully digital I&C nuclear reactor. The PUR-1 DT is a digital platform for, amongst others, transfer learning experimentation (Theos, 2023). The platform has AI/ML modules trained from the bottom up, first, learning about different sensors (from their own data), then move up to component levels, functions, tasks, and eventually to subsystem and system levels. Figure 6.20 illustrates an instance of horizontal transfer learning, where a model built on data from sensor 1, can be transferred, or reused, on sensor 2. The model is tested on sensor 2. If the test is passed, then the model is given a "badge," a certificate of competence along with the passing score (say 89/100). In this way the knowledge embodied in the model of sensor 1 can be transferred to sensor 2, if and only if, certain conditions are met; e.g., if the passing score is 85 or higher. The AI/ML modules may teach each other and get qualified through transfer learning protocols that lead to acquisition of quantitative credentials for skills and qualifications needed for maintaining functionality at prespecified levels of reliability under all conditions.

Sensing equipment in nuclear facilities may be prone to degradation and failure due to aging, harsh temperatures, and radiation conditions resulting in downtime and large operations and maintenance (O&M) costs. Automation of sensor monitoring, if implemented correctly, can reduce O&M costs using AI/ML modules with demonstrated competence in various scientific and engineering fields, for example, computer vision; and brought to bear in the nuclear context

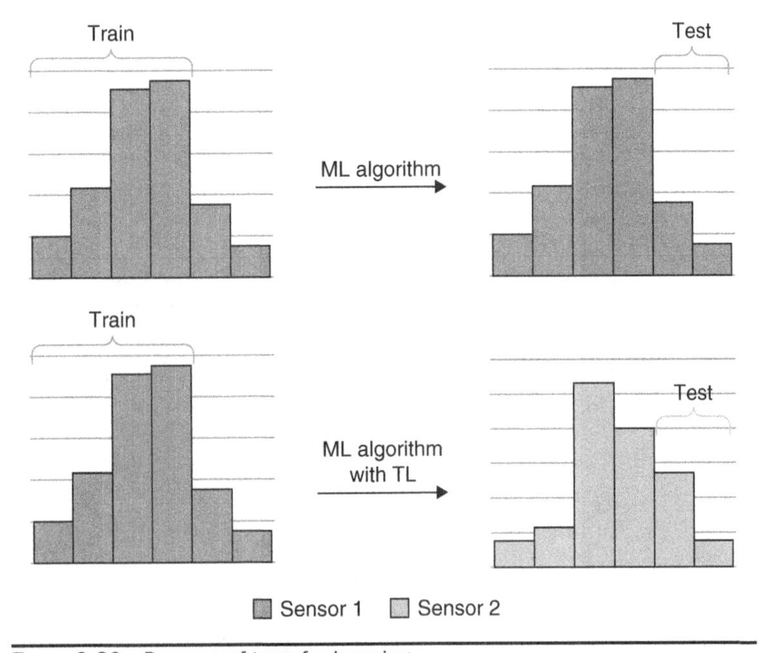

Figure 6.20 Process of transfer learning.

through transfer learning. An advantage of AI/ML modules is that they do not rely on detailed knowledge of a reactor system, which might be challenging to acquire because of uncertainty in loss terms in the experimental system, uncertainty in material property values at high temperatures, and administrative restrictions due to regulatory requirements or confidentiality and privacy concerns. Instead, AI/ML modules can construct a model based on input and output values of historical sensor data, enabling AI/ML to predict sensor readings during the monitoring process.

Transfer learning may be used for *soft extrapolation learning*, which refers to the ability of a model to apply what has already been learned to new scenarios or data points that have not been seen before. This feature is advantageous for nuclear applications, where physics and engineering experiments are expensive or may be conducted rarely, and the amount of data available to train new models is limited. However, this method can be challenging, as AI/ML modules may struggle to generalize to situations that are significantly different from their training data. This can lead to problems such as *overfitting*, where a model becomes too specialized to the training data and fails to generalize to new data and therefore becomes untrustworthy. Another type of problem is *underfitting*, where a model fails to capture the underlying patterns in the data and performs poorly on both the training and testing datasets leading again to untrustworthiness. To address these and similar challenges, AI/ML modules may include *pretrained models* functioning as an initial option for a new task or problem. Instead of training a module from scratch on a new task, the knowledge gained from a previously trained model (often on a somewhat related task or dataset) is transferred to the new model, while fine-tuning is performed to adapt the model to the specific task at hand. The primary goal of transfer learning in most applications is to leverage knowledge acquired from solving one task or dataset, to improve the performance of a module applied to a different but related task or dataset (Pantopoulou M, 2023). Additionally, transfer learning must ensure the existence of significant "ratcheting" in quantifiable ways. AI/ML models that leverage transfer learning using real data from PUR-1 are integrated into PUR-1 DT to achieve accurate sensor monitoring under different reactor transients.

AI/ML modules such as *recurrent neural networks* (RNN) can be used to validate sensor readings, which can provide certain benefits such as optimal planning and time to minimize forced shutdowns. Recurrent neural networks are distinguished by their memory, as they obtain information from prior inputs to affect current input and output. While conventional deep learning (statistical) modules operate on the hypothesis that inputs and outputs are independent of each other, the output of RNNs depends on prior elements of the sequence. Long short-term memory (LSTM) and gated recurrent unit (GRU) networks can be used for sensor monitoring purposes (Pantopoulou S, 2022). These models are specific types of *recurrent neural networks*, which work to address the problem of long-term dependencies; that is, their structure contains hidden states that depend on previous ones. Another algorithm that can be used for sensor monitoring is the *auto-regressive integrated moving average* (ARIMA) model and its variations. The process of developing an ARIMA model includes a preprocessing phase of the dataset, which allows it to perform better, as it is specialized on the examined dataset. ARIMA is a statistical analysis model which is used to predict future trends, and performs well with stationary time series data with a clear pattern. For cases of nonstationary or highly complex data, more advanced time series models such as seasonal ARIMA (SARIMA) or other ML algorithms can be used. Furthermore, physics-informed neural networks (PINNs) are highly effective for neutron/flux monitoring (Prantikos, 2022). PINNs encode fundamental physical laws, alleviating the need for reliance on big data availability for model training. These models leverage automatic differentiation, enabling accurate and rapid computations, which is highly promising for real-time applications.

As discussed in the fuzzy control section (Sec. 6.2), the nuclear reactor PUR-1 is an all-digital 10 kWth material test reactor (MTR)—pool type, with fuel consisting of high assay low enriched uranium (19.75% ^{235}U) in the form of U_3Si_2—Al. There are total 16 assemblies, where each standard assembly has up to 14 fuel elements. The core is submerged into a 5.2-m-deep water pool, where water is used for both neutron moderation and fuel heat removal. Average thermal neutron flux in the fuel region is 1.2×10^{10} n/cm^2·s, with the maximum thermal flux reaching the value of 2.1×10^{11} n/cm^2·s. The reactor power is controlled with three control rods. Two of them are borated stainless steel shim safety rods (SS1 and SS2), and the third one is 304 stainless steel regulating rod (RR). The PUR-1 DT receives signals from reactor sensors in real-time, performs diagnostics and predictive modeling using physics-based models, and then sends commands back to a remotely auxiliary control rod to achieve a predetermined objective, for example, load following. The PUR-1 DT was developed by Purdue's Radiation Imaging and Nuclear Sensing Laboratory directed by professor Stylianos Chatzidakis. A schematic of the PUR-1 DT is shown in Fig. 6.21.

The instrumentation and control system of PUR-1 allows for real-time monitoring of more than 2000 signals including sensor, operational, and network data using a graphical user interface (GUI). Through the GUI several variables are monitored in real time, including (i) control rod motion and position, (ii) temperature, flow rate, radiation, and neutron flux sensor data, (iii) current and future predictions of unobserved parameters including state identification and fuel and coolant temperatures, and (iv) performance metrics and data analytics. The data, including sensor signals, commands, and performance metrics,

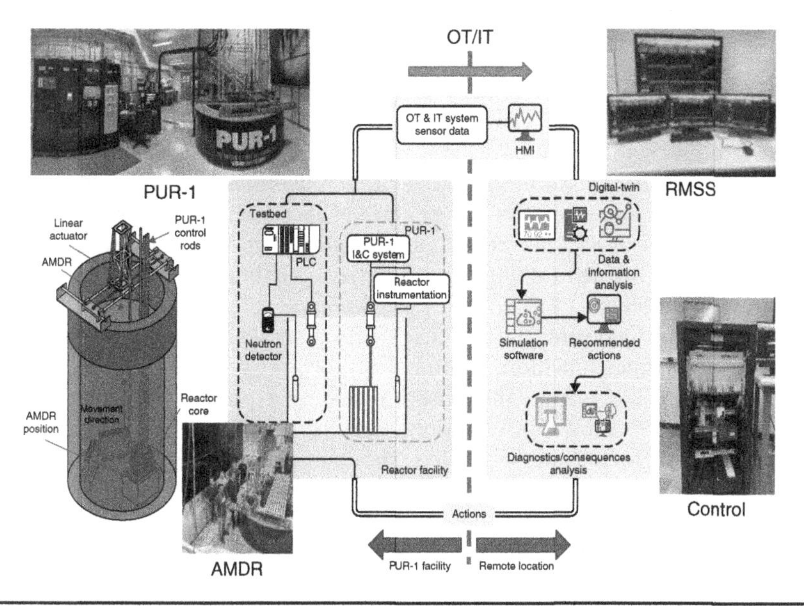

Figure 6.21 Schematic of PUR-1 and its digital twin, PUR-1 DT.

are sent to a dedicated data storage unit with millisecond resolution to allow easy retrieval for data analytics and performance evaluation.

How does a system like PUR-1 DT know when transfer learning is called for and can this decision be found in the data alone? An important element in addressing these and similar questions is identifying the emergence of some significant change in the system that necessitates new learning. We can think of this significant change as the emergence of a trend. Identifying the onset of a trend, on data alone, is key to transfer learning. A significant shift in the constitutive relations underlying a system calls for new models to be brought to bear on mirroring its new configuration and transfer learning to integrate and fine-tune its DT. The question of whether a new configuration is emerging (from the data alone) and how to detect it automatically, is an important question in data analytics and big data and finds application in various fields, including business, finance, technology, and social sciences. In digital systems of instrumentation and control (I&C) the onset of a trend may quickly be obfuscated due to compensatory or corrective responses of controllers. Hence, trend identification is crucial. Conventional analytics rely on statistical approaches and focus on rather static data analytics, that is, they assume the system is in steady state, there are no major transitions and no new models are called for. In many cases, the monitoring system ought to provide timely and reliable response to incoming data (from online signals), that is, a trustworthy decision needs to be made in a relatively small window of time. Although conventional methods are not appropriate for these tasks, fuzzy logic can make decisions based on data alone and, most importantly, recent data. Essentially, it focuses on features which have the data embedded in them. Thus, the method is signal- and process-independent, and hence, can be easily used with *transfer learning* to other processes and systems. This was advanced by Dr. Xin Wang as part of his doctoral dissertation and is proven useful in automatically detecting the onset of trends (Wang, 2003). Each individual trend is identified as *increasing, decreasing*, or *constant* based on six statistical variables, which form universes of discourse enabling a quick and accurate decision.

To classify incoming data of a DT into *increasing, decreasing*, or *constant* in a timely fashion, the most important features from the data are extracted within the time interval of a prespecified size, called the current time window. Altogether six variables representing several useful features are extracted from current and adjacent data points. None of these features is dominant in making a final decision. They need to be fused through a fuzzy logic algorithm. The six variables use as values triangular fuzzy numbers, which are fused to produce a final fuzzy number and a trend decision is made. The flowchart for the approach is shown in Fig. 6.22.

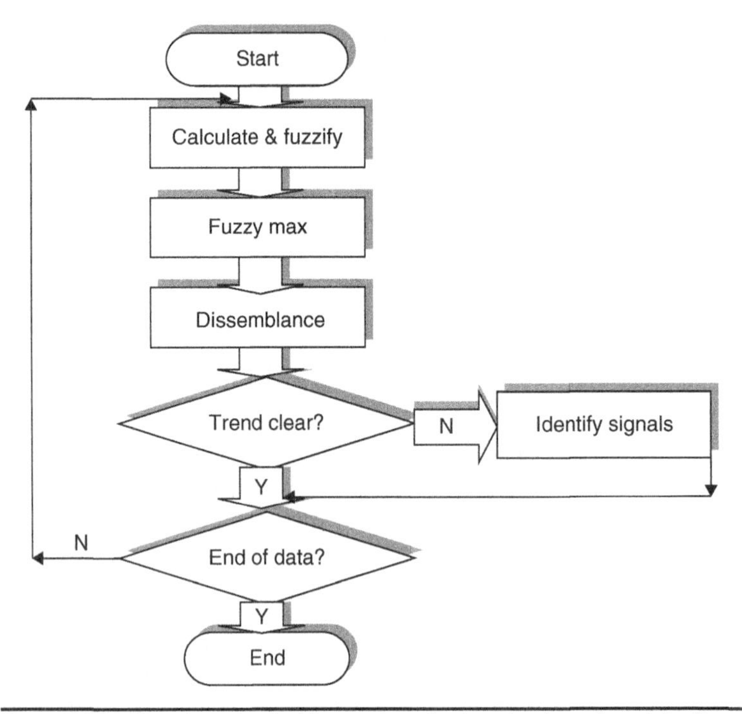

Figure 6.22 Flowchart for deciding on transfer learning based on detecting the emergence of a trend.

The information on the emergence of a new trend is contained within the six features associated with each data space, essentially the outputs of instruments. The six features depend on the sample points within a short time window near the present. The features include important properties of the data, such as deviation from steady-state and time rate of change, as elucidated next.

From the analytics of sensor data, we know that additive noise is approximately Gaussian and distributed with mean equal to zero and some standard deviation σ. Signals corrupted by noise are Gaussian random processes with mean $\mu(t)$ and standard deviation σ. Whenever a transient occurs, the mean of the signal changes from $\mu(t_1)$ to $\mu(t_2)$, although the standard deviation remains the same. Hence the decision as whether or not there is a significant change reduces to the mathematical problem of distinguishing between two Gaussian distributions at the onset of a transition and as fast as possible. In order to summarize the information, six corresponding fuzzy numbers are developed for each data space (from a signal). The physical meaning and mathematical description of fuzzy numbers are as follows:

Probability Density Function (*pdf*)

The probability density function (*pdf*) is based on the following definition:

$$pdf_{t_c} = \frac{1}{\sqrt{2\pi}\sigma} \exp^{-\frac{(s_{t_c}-\mu_s)^2}{2\sigma^2}} \tag{6.7}$$

where,

t_c = current time step
S_{t_c} = signal value at t_c
μ_s = mean value of the steady-state signal
σ = standard deviation of the steady-state signal

Sample points belonging to off-normal states are far from the steady-state mean and hence have small *pdf* values. Thus, this variable shows the deviation of the signal from steady-state operation. Only the current data value is needed to compute this variable. Because the additive noise can affect the variable significantly, it is not a "reliable" feature and for this reason it should be given a smaller importance in the overall decision process.

Cumulative Probability Density Function (*cum_pdf*)

The cumulative probability density function (*cum_pdf*) is defined as follows:

$$cum\_pdf_{t_c} = \sum_{i=0}^{n} pdf_{t_{c-i}} \tag{6.8}$$

where, n represents the length of the time window.

Small signal changes are accumulated and recorded in this variable. By accumulating small signal changes, it becomes possible to make decisions based not only on the instantaneous signal changes, for example, the *pdf*, but also on the recent history of the signal. Due to its cumulative property, it is a very important feature in the decision strategy.

Probability Density Function of Average Value (*pdfofavg*)

The probability density function of average value (*pdfofavg*) can be represented as follows:

$$pdfofavg = \frac{1}{\sqrt{2\pi\sigma}}\exp\left(-\frac{(avg-\mu)^2}{2\sigma^2}\right) \tag{6.9}$$

where,

$$avg = \left(\sum_{k=0}^{4} s_{t-k}\right) / 5$$

μ_s = mean value of a steady-state signal

σ = standard deviation of a steady-state signal

S_{t-k} = the sample value k time steps before the current point

If the value of this variable is very small, there is significant confidence for a positive decision regarding the emergence of a trend. This feature also possesses cumulative characteristics. It is most helpful in distinguishing two Gaussian distributions by reducing their respective standard deviations.

Average Derivative (*avgd*)

The definition of average derivative (*avgd*) is shown below:

$$avgd_{t_c} = \left|\frac{wavg_{t_c} - wavg_{t_{c-1}}}{t_c - t_{c-1}}\right| \tag{6.10}$$

where,

$$wavg_{t_c} = \frac{\sum_{i=0}^{n}\left[e^{-k(t_c - t_{c-i})} s_{t_{c-i}}\right]}{\sum_{i=0}^{n} e^{-k(t_c - t_{c-i})}} \tag{6.11}$$

This variable represents the time rate of change of the variable *wavg*, which is defined as the weighted sum of the sampled signal values over the time window of length n. In Eq. (6.11), k is a positive constant. The larger the value of this variable, the more likely that a significant change (or a transient) is starting. Because even in steady-state, this variable can have large quantity due to noise, it is not an important feature in the decision strategy.

Relative Deviation (*ravg*)

The variable *relative deviation* (*ravg*) is based on the following definition:

$$ravg_{t_c} = \left| \frac{avg_{t_c} - \mu_s}{\mu_s} \right| \tag{6.12}$$

where,

μ_s = mean value of a steady-state signal

This variable represents the deviation of *avg* from the signal mean steady-state value and is independent of the amplitude of the signal. Because of its cumulative property, it can play an important role.

Sample Derivative (*sd*)

The definition of the variable *sd* is shown below:

$$sd_{t_c} = \left| \frac{s_{t_c} - s_{t_{c-1}}}{t_c - t_{c-1}} \right| \tag{6.13}$$

This variable is used to capture the instantaneous rate of change of the original signal without any smoothing since smoothing hides the occurrence of local peaks.

Developing a general index to identify the onset of a significant change (a new trend or transient) that is departure of the most recent data point (*truth*) from equilibrium, the six variables are fuzzily categorized to achieve a decision on transfer learning. Since the six variables have their own values, each significantly different from the others, it is difficult to integrate the information unless it is transformed into a common scale. A fuzzy variable named *truth* is created to represent the truthfulness of deviation from equilibrium, for each variable. For example, after the value of one variable is obtained, the deviation of the value from equilibrium is considered according to the value and a variable *truth* in the interval [–1, 1]. As the deviation grows in magnitude, so does the absolute value of *truth*, implying that the distance of the current data from equilibrium has increased. The sign of the *truth* value indicates whether the deviation is positive or negative. For example, if a feature has *truth* of 0, it is surmised that this feature shows that the current sample point belongs to a *steady-state* or *equilibrium* signal. On the other hand, if a feature has *truth* of –1, it means the feature shows the current sample point belongs to a *decreasing* trend. Altogether six *truth* decision curves are set up to implement the transformation process. Each variable has its own *truth* decision curve which is derived from a Gaussian distribution curve.

Based on the *truth* of each feature, six triangular fuzzy numbers are developed whose membership corresponding to *truth* is maximum 1, and corresponding to 0, 1, or −1 is 0. The process is shown in Fig. 6.23.

The final fuzzy number indicating the onset of a new situation and the need for transfer learning is obtained through a max operation. For the convenience of summarizing all the trend information contained in the features, a final fuzzy triangular number is computed with the "max" operation. The final decision is made on the basis of this new fuzzy number.

$$(a^{\alpha}, b^{\alpha}) = \left(\max_{i=1\sim5}(a_i^{\alpha}), \max_{i=1\sim5}(b_i^{\alpha})\right) \tag{6.14}$$

where,

a^{α} = the left-end-point of the of α-cut of the final fuzzy number
b^{α} = the right-end-point of the of α-cut of the final fuzzy number
a_i^{α} = the left-end-point of the of α-cut of the *i*th fuzzy number
b_i^{α} = the right-end-point of the α-cut of the *i*th fuzzy number

It should be recalled from Sec. 2.4 that α-cuts are intervals having left- and right-end-points for each level determined by the parameter α. Obtaining the final fuzzy number to make a decision calls for defuzzification. The distance between the final fuzzy number

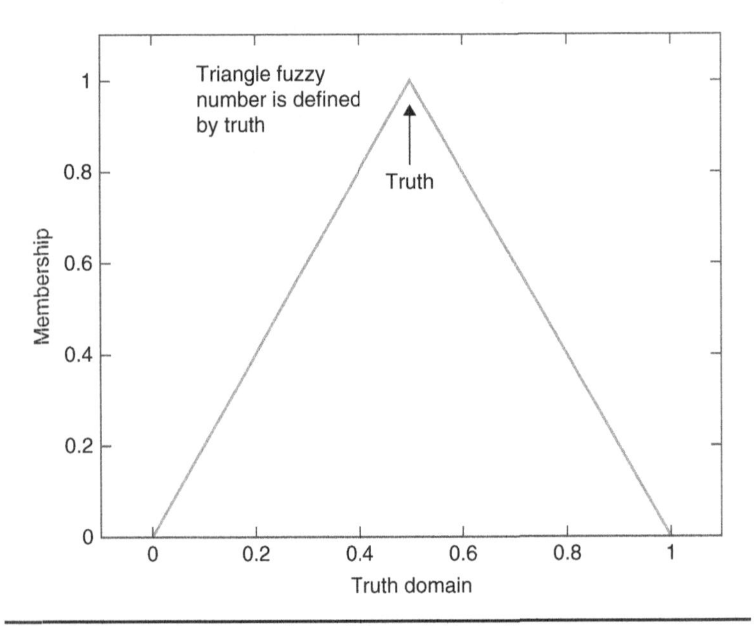

FIGURE 6.23 Setting fuzzy values of the variable truth.

and prototype membership functions is calculated and compared in this process. The distance is also called the *dissemblance index* (*DI*) of two fuzzy numbers A and B, $\delta(A, B)$, and is defined as:

$$\begin{aligned}\delta(A,B) &= \int_{\alpha=0}^{1} \delta(A_\alpha, B_\alpha)\,d\alpha \\ &= \frac{1}{2}\beta \int_{\alpha=0}^{1} \Delta(A_\alpha, B_\alpha)\,d\alpha \\ &= \frac{1}{2}\beta \int_{\alpha=0}^{1} \left(\left|a_1^\alpha - b_1^\alpha\right| + \left|a_2^\alpha - b_2^\alpha\right|\right)d\alpha\end{aligned} \quad (6.15)$$

where,

β is used to normalize the value of DI to [0, 1]

a_1^α (b_1^α) = the left-end-point of the α-cut of a fuzzy number of A (B)

a_2^α (b_2^α) = the right-end-point of the α-cut of a fuzzy number A (B)

$\Delta(A_\alpha, B_\alpha)$ = distance between α-cuts of fuzzy numbers A and B (Wang, 2003)

The prototype membership functions shown in Fig. 6.24 are fuzzy numbers 0, +1, and −1, which represent *steady-state, increasing, decreasing* transients, respectively. The distances between the final fuzzy number and the four prototype fuzzy numbers are compared and the closest prototype is chosen as the result of the decision process.

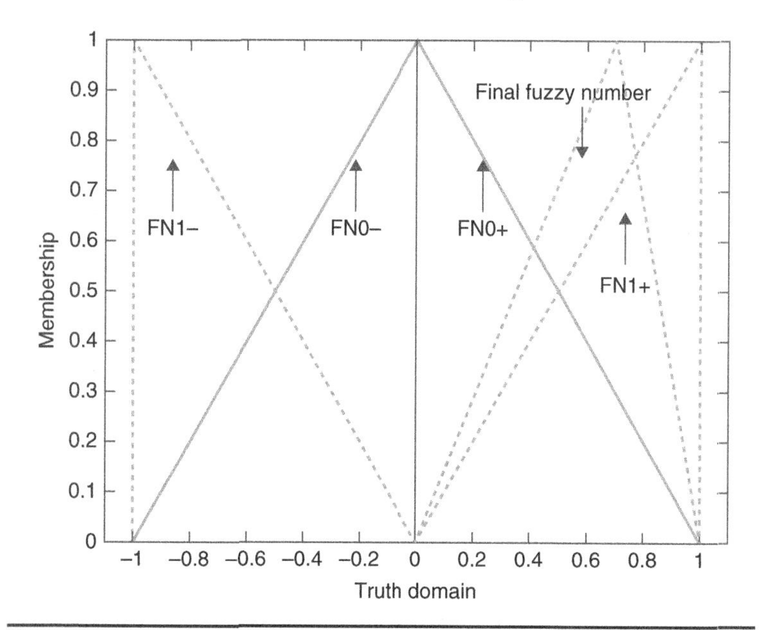

FIGURE 6.24 Four prototype fuzzy values categorizing the onset of the need for transfer learning.

A variable *confid* is introduced to quantify the confidence of the decision. It is defined as:

$$confid = \frac{\delta_{1+,1-}}{\delta_{0+,0-} + \delta_{1+,1-}} \tag{6.16}$$

where,

$\delta_{1+,1-}$ = distance or *DI* from the final fuzzy number to fuzzy 1+ or fuzzy 1–

$\delta_{0+,0-}$ = distance or *DI* from the final fuzzy number to fuzzy 0+ or fuzzy 0–

This variable is helpful for us to improve a fuzzy-based decision strategy. The unique threshold which is used frequently in nonfuzzy decision strategy is avoided here. One major characteristic is that the decisions made are not only dependent on the values of $\delta_{0+,0-}$, $\delta_{1+,1-}$ and *confid*, but also on the previous values of *confid*. Actually, in many cases, the changing trend of *confid* is even more important than the variable itself. It is validated that this strategy is very useful in determining to which state the current sample point pertains and contributes to quicker response and more stable results.

Based on this fuzzy methodology numerous tests with various data have been performed to validate the effectiveness of the approach. Consider for example, Figs. 6.25 and 6.26 which show two flow signals

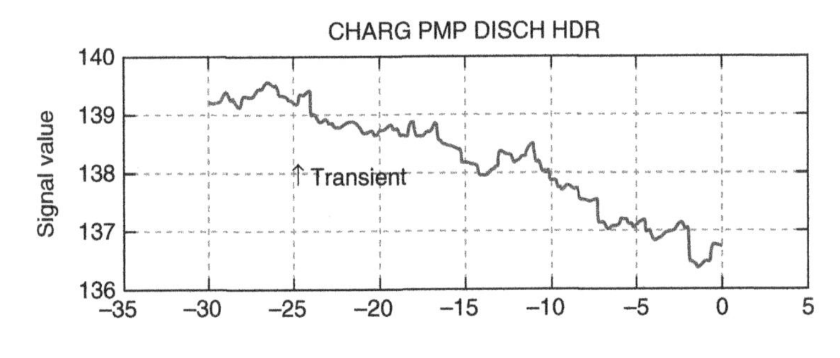

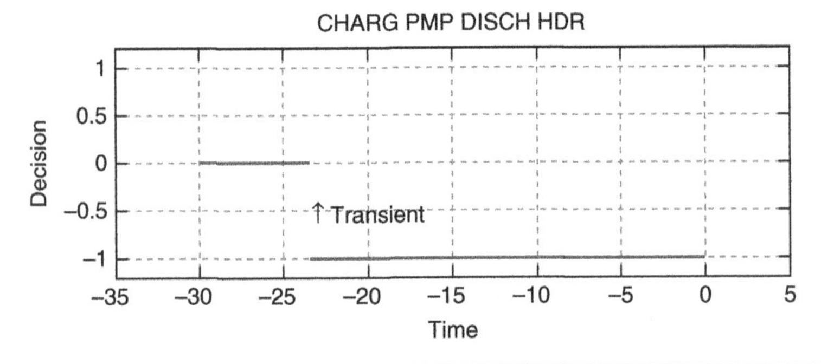

Figure 6.25 A signal (upper) at the 24th time step shows that a trend emerges and the tool (below) identifies that there is at that point decreasing trend emerging.

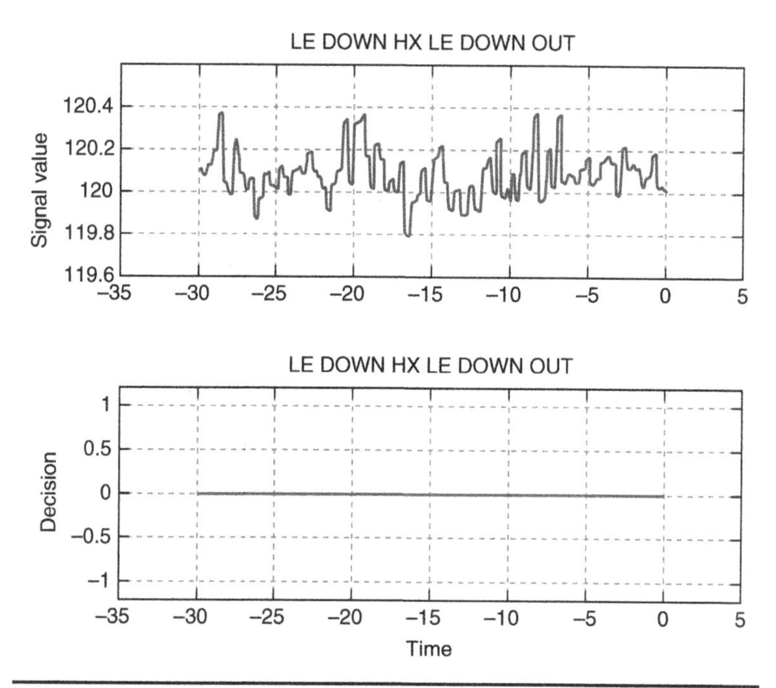

Figure 6.26 The fuzzy tool decides that there is no change (lower) even though the visual information of the signal (upper) may suggest otherwise.

featuring measurement data from the chemical and volume control system (CVCS) of a pressurized water reactor. The flow measurements are sampled at 5-sec intervals. Figure 6.25 illustrates the changing pump discharge header flow and Fig. 6.26 illustrates the outlet flow of the letdown heat exchanger.

It should be noted that the signals shown in Figs. 6.25 and 6.26 are transient and steady-state, respectively and that the onset of the transient is identified 10 sec after onset, when the change is only about 0.3%. For the steady-state situation, the response is correct despite all the fluctuations and the noise in the signal.

To decide when transfer learning needs to take place based on data alone, one can perform online estimation of trends through the six statistical variables defined and transformed into a final fuzzy number. The trend identification is based on the data alone and is a quick but accurate decision strategy which is especially helpful when the signal data represents a small or slow transient, as shown in numerous validation tests (Wang, 2003).

6.7 Medical AI with Deontic and Fuzzy Logic

Medical AI is concerned with questions of health and treatment of disease. It is one of the early areas of AI going back to the 1980s, when now famous medical expert systems such as MYCIN offered diagnostic advice based on symptoms and clinical data. Over the past several decades, medical applications have expanded beyond expert systems to include modules that perform sophisticated diagnoses, treatment protocol development, drug design, patient monitoring, sophisticated anatomical visualizations, and personalized healthcare that may also feature counseling and mental-health guidance and therapy.

The practice of medicine is a highly regulated activity and as such it calls for *deontic logic*. This is a special branch of logic that focuses on the "*ought*," "*should*," and "*should-not*" aspects of reasoning (Daskalopulu, 2022). In contrast to fuzzy logic which focuses on parsimony *vis-á-vis* flexible categorization (fuzzy sets) within the Chomskian notion of "discrete infinity," deontic logic deals with crisp "ought" and "ought not" statements, which are related to norms, permissions, obligations, and prohibitions. Deontic logic is typically used to encode relationships between norms, actions, and their logical implications, such as "*it is obligatory to do X*" or "*it is forbidden to do Y*" where *X* and *Y* are representations of action, which can be construed as simple or compound qualitative facts, often without grades of truth or degrees of uncertainty, which may increase the computational cost to astronomical heights.

Fuzzy logic has been used extensively in medical AI. For instance, considering disease through the normativistic perspective of Sadegh-Zadeh, an action approach views health-related knowledge of an individual through data structures that include, but are not limited to, medical and social records, disease diagnoses, prognoses, and anticipations of health and wellness throughout the life of a person (Sadegh-Zadeh, 2000). Fuzzy logic enters as means of managing the complexity of decision-making by sacrificing some unnecessary precision to protect the overall accuracy of the process and fit it to the specific patient. Normative and fuzzy aspects of medical AI focus on tracking lifelong evolution of data and knowledge about an individual together with ethics and regulatory compliance regarding privacy, cybersecurity, and the like. In this framework, fuzzy logic and deontic logic can work together to produce approaches that consider disease as a complex social artifact mapped onto deontic constructs and fuzzy logic.

In medical AI, variables reflecting the state of health or deviation from a normal state of health may be used in place of *error, change in error,* or *history of error*, typically used in control (see Chap. 4). The medical variables may depend on the availability of parameter and structure estimation knowledge and may not be directly measurable variables, for example, *wellness*, or *healing*, or variables which can be estimated through modules that operate on demand. As in control language-games, any directly manipulated fuzzy variable in a medical set of norms encoded as *if/then* rules, will reside in the *RHS* of rules. The output is a decision leading to action such as drug type and dose or recommendation on appropriate treatment protocol.

The consensus view on medical AI identifies two unique and fundamental characteristics: the normative nature of medical practice and the fuzzy nature of the information that it employs.

The first relies on rules of protocols, clinical research, diagnosis, and disease treatment. These rules are *normative*, they describe reference states against which actual clinical states are compared to diagnose or determine the efficacy of medical practice regarding professional, ethical, and regulatory compliance. Through this perspective medical AI evaluates the "distance" between actual and ideal states; drug administration and dosage monitoring systems follow actual and (ideally) anticipated patient progress from disease to health. To bridge the gap between "ideality" to "actuality" deontic and fuzzy logic enable more nuanced distinctions between "healthy" vs "unhealthy" in a variety of contexts as formulated by professor A. Daskalopulu.

The second characteristic recognizes that medical practice can be extraordinarily complex and its application highly idiosyncratic, hence practice norms are best to be considered as fuzzy language-games. For instance, a cough that is *severe* according to a patient might be seen as *normal* by a doctor. In norms pertaining to access to data or the specification of hospital protocols for certain procedures we often encounter terms such as "interested party" or "qualified personnel" or "appropriate procedure" or various temporal references such as "within reasonable time." These are notions that are not strictly defined and may be interpreted subjectively. The synergisms of fuzzy logic and deontic logic results in tools with multimodal and advanced dialogue capacities.

For instance, the core concepts of deontic logic and fuzzy logic may be brought to bear on novel synergisms as illustrated below for cases where they are used jointly to model scenarios that may be typical in medical practice.

In deontic logic, the necessity operator corresponds to *O* (bligation), or *"it ought to be the case that"* statements, and the possibility operator corresponds to *P* (ermission), or *"it may be the case that"* statements. Standard (propositional) deontic logic (*SDL*) is the branch of deontic logic most commonly used. In *SDL*, the deontic operators *O* and *P* are *interdefined*, that is, they describe an action as permitted if its complement is not obligatory,

$$P\alpha \equiv \sim O \sim \alpha \tag{6.17}$$

where, $P\alpha$ is permitted action, and,

$\sim O$ is the negation of the obligation operator acting on the negation of action α

The sign $\sim$ is used in *SDL* as the symbol for negation. Similarly, a derived operator is defined for *F* (orbiddance), where something is forbidden if its complement is not permitted and likewise something is forbidden if its complement is obligatory. Symbolically this is indicated as

$$F\alpha \equiv \sim P \sim a \qquad (6.18)$$

or equivalently,

$$F\alpha \equiv O \sim \alpha \qquad (6.19)$$

where, the symbols found in Eqs. (6.18) and (6.19) are as in Eq. (6.7).

Two points need to be examined, both pertaining to the use of deontic logic for the representation of medical norms. The first regards the syntax level of descriptions as is seen in Eqs. (6.17), (6.18), and (6.19). A deontic operator might operate on a state or on an action. In the former case an expression of the form $O\sigma$ is to be understood as *"it ought to be the case that* σ *holds"* and in the latter an expression of the form $O\alpha$ is to be understood as *"it ought to be the case that* α *is performed."* Such cases use language-games for norms and procedures of the form *"it is obligatory that the doctor performs a test"* or *"it is forbidden that a drug dosage exceeds a certain amount."* It should be noted here that the logical descriptions denoting either a state or an action are syntactically indistinguishable. In order to account for both cases where specific actions are deontically qualified explicitly and for cases where particular means are rather specific, the *SDL* needs to be extended to include an action operator $E_x S$ with the intended reading *"agent x sees to it that S is the case"* or *"agent x brings about S."* In this way, explicit agency can be assigned for the accomplishment of states. If a particular state *S* is achieved by one or more specific actions $\alpha_1, \alpha_2, \ldots$ such information can be reflected in an appropriate definition of the action operator via implication rules, in effect creating a meta-level where it is determined, possibly through a cluster of fuzzy *if/then* rules, whether the performance of a specific action by a specific agent counts as a valid instance resulting in a certain state of affairs.

The second regards the idiosyncratic nature of medical norms, which are not impersonal, that is, in addition to specifying what is obliged, permitted or forbidden, one needs to distinguish the norm bearer from the actor. Medical norms might appear impersonal and the process of constructing a deontic logic representation for them is helpful in establishing whether such an omission is intentional (should questions about norm violation arise let the inquiry decide whose fault it was) or inadvertent, and in the latter case questions about roles and responsibility distribution may be probed.

To illustrate these points and the issues of representation, consider the following *if/then* rule encountered in medical practice:

> *If a patient complains of acute chest pain that radiates to her left arm, then, if you want to know whether she has myocardial infarction, [then] it is obligatory that you record an ECG or determine the concentration of heart-relevant enzymes in her blood.*

This compound rule can be translated into an abstract *SDL* form, if the following breakdown of its parts is made:

CP(x): x complains of acute chest pain that radiates to x's left arm;
DR(y, x): y is the doctor of x;
ECG(x): an ECG is recorded for x;
ENZ(x): the concentration of heart-relevant enzymes in x's blood is determined;
WK(y,x): y wants to know something about x.

Then, an *SDL* formulation (translation) of the rule is:

$$(CP(x) \rightarrow (DR(y,x) \wedge WK(y,x)) \rightarrow (ECG(x) \vee ENZ(X))) \qquad (6.20)$$

Contrast Eq. (6.20), with a semantically equivalent expression that can be found in medical AI, which is the following:

$$((CP(x) \wedge DR(y,x)) \rightarrow O_y(E_yECG(x) \vee E_yENZ(x)) \qquad (6.21)$$

Equations (6.20) and (6.21) are syntactically different (they look very different) but are indeed semantically equivalent, that is, they say the same thing. Hence their use is limited. An additional conditional (a fuzzy *if/then* rule) relates the particular *ECG* and *ENZ* facts to the diagnosis of myocardial infarction for the patient.

It follows from the example that logical formulations such as Eqs. (6.20) and (6.21) are limited also in the sense that they cannot distinguish between what the physician ought to do and what he actually does, in response to a patient complaint. The *normative character* of the norm is conflated to a *declarative statement*. Should a complaint arise it is difficult to establish who is responsible for the nonperformance of the tests specified in the norm. For instance, an *ECG* test for a patient may entail a series of specific actions (such as turning the *ECG* machine on, connecting the patient to the machine via the appropriate electrodes, instructing the patient to lie still, etc.). By using the representation $E_yECG(x)$ to denote that "*y sees to it that an ECG is recorded for x*" we essentially specify the desired outcomes rather than the individual actions required to achieve it.

As another example, consider the growing interest in genetic testing to determine, for example, what is the risk of a patient developing prostate cancer or pancreatic or stomach malignancies; breast and ovarian cancers in women (now easily performed through checks for mutations of BRCA and a growing set of other genes). Not only can such risk be determined, but moreover the likely age of onset of such cancers and the survival chances of the patient may be estimated, too. Interesting questions of a normative nature arise in relation to these data including, but not limited to, what are patient rights for access to data; does the patient have to share the data with an insurance company; should any medical practitioner other than an oncologist, say an ophthalmologist, have access to the data; what are the obligations, rights, privileges of companies or state institutions toward the patient. Obviously legal notions of duty, right, privilege, and complex normative concepts such as authority, liability, and immunity enter the analysis of such problems and fuzzy language-games ought to be formulated for tools that answer questions of liability and enforce appropriate reparatory mechanisms while respecting human rights and freedoms. Although there appears to be an antithetical relation between fuzziness and norms, in actuality, there is a natural synergism between the two.

The above examples illustrate the possibilities of fuzzy logic augmenting other logical representations. Fuzzy algorithms may offer supervision, clarifications, explanatory facilities, and transfer learning that may allow a standard deontic logic representation to achieve greater functionality. They may easily capture various shades of agency and responsibility which could be useful when actual events and states of affairs are compared against norms. In this manner, context-specific needs are recognized and a medical AI system determines whether the norms are satisfied or violated and by whom under full respect for human rights and freedoms and through the application of trust principles to technology.

References

Appiah, R., Oktavian, M. R., and Tsoukalas, L. H., "Neuro-Fuzzy Predictive Power Controller for a Nuclear Research Reactor," *Proceedings of the 2022 American Nuclear Society*, Phoenix, AZ, November 13–17, 2022.

Daskalopulu, A., Tsoukalas, L. H., Bargiotas, D., "Normative and Fuzzy Components of Medical AI Applications," in Virvou, M., Tsihrintzis, G. A., Bourbakis, N. G., and Jain, L. C. (eds.), *Handbook on Artificial Intelligence-Empowered Applied Software Engineering. Artificial Intelligence-Enhanced Software and Systems Engineering*, Vol. 3. Springer, 2022.

Lamarsh, J. R., and Baratta, A. J., *Introduction to Nuclear Engineering*, 3rd edition, Prentice Hall, 2001.

Mertoguno, J. S., Craven, R. M., Mickelson, M. S., and Koller, D. P., "A Physics-Based Strategy for Cyber Resilience of CPS," in *Autonomous Systems: Sensors, Processing, and Security for Vehicles and Infrastructure 2019*, International Society for Optics and Photonics, Vol. 11009, 2019.

Octavian, M. R., Appiah, R., Lastres, O., Miller, T., Chapman, A., and Tsoukalas, L. H., "Fuzzy Power Controller Design for Purdue University Research Reactor-1," *Proceedings of the International Conference on Computational Nuclear Engineering and Numerical Methods*, New York, June 2022.

Pantazopoulos, K. N., Tsoukalas, L. H., Bourbakis, N. G., Brun, M. J., and Houstis, E. N., "Financial Prediction and Trading Strategies using Neurofuzzy Approaches," *IEEE Transactions on Systems, Man, and Cybernetics, Part B (Cybernetics)*, Vol. 28, No. 4, pp. 520–531, August 1998, doi: 10.1109/3477.704291.

Pantopoulou, M., Kultgen, D., Tsoukalas, L. H., and Heifetz, A., "Machine Learning-Based Monitoring of Liquid Sodium Vessel Heater Zones," *Transactions of the ANS 2022 Winter Meeting*, Vol. 127, No. 1, pp. 308–311, 2022.

Pantopoulou, S., Ankel, V., Weathered, M. T., Lisowski, D. D., Cilliers, A., Tsoukalas, L. H., and Heifetz, A., "Monitoring of Temperature Measurements for Different Flow Regimes in Water and Galinstan with Long Short-Term Memory Networks and Transfer Learning of Sensors," *Computation*, Vol. 10, No. 7, p. 108, 2022.

Pantopoulou, S., *Cybersecurity in the PUR-1 Nuclear Reactor*, MSNE Purdue University Thesis, 2021.

Prantikos, K., *Physics-Informed Neural Network Solution of Point Kinetics Equations for PUR-1 Digital Twin*, MSNE Thesis, Purdue University, 2022.

Prantikos, K., Tsoukalas, L. H., and Heifetz, A., "Physics-Informed Neural Network Solution of Point Kinetics Equations for Development of Small Modular Reactor Digital Twin," in *Proceedings of the 2022 American Nuclear Society*, Anaheim, June 12–16, 2022.

Sadegh-Zadeh, K., "Fuzzy Health, Illness and Disease," *Journal of Medicine and Philosophy. Forum Bioethics and Philosophy of Medicine*, Vol. 25, No. 5, pp. 605–638, 2000.

Salazar, W. C., Machado, D. O., Len, A. J. G., Gonzalez, J. M. E., Alba, C. B., de Andrade, G. A., and Normey-Rico, J. E., "Neuro-Fuzzy Digital Twin of a High Temperature Generator," *IFAC-PapersOnLine*, 55, pp. 466–471, 2022. doi:10.1016/j.ifacol.2022.07.081.

Theos, V., Gkouliaras, K., Dahm, Z., Miller, T., Jowers, B., and Chatzidakis, S., "A Physical Testbed for Nuclear Cybersecurity Research," *Transactions of the ANS 2023 Annual Meeting*, Vol. 128, No. 1, pp. 175–178, 2023.

Townsend, C. H., *Technical Specifications for the Purdue Reactor-1 Docket Number 50-182*. West Lafayette, IN, 2016.

Turing, A. M., "Computing Machinery and Intelligence," *Mind*, Vol. 49, pp. 433–460, 1950.

Wang, X., *A Feature-Based Transient Signal Detection Methodology*, Ph.D. Dissertation, Purdue University, 2003.

Weiss, K., Khoshgoftaar, T. M., and Wang, D., "A survey of Transfer Learning," *Journal of Big Data*, Vol. 3, No. 1, pp. 1–40, 2016.

Zadeh, L. A., "PRUF—A Meaning Representation Language," *Fuzzy Reasoning and Its Applications*, E. H. Mamdani and B. R. Gaines (eds), *Computer and People Series*, pp. 1–66, Academic Press, 1981.

APPENDIX A
Python Script for Example 4.1

```
import numpy as np
import skfuzzy as fuzz
import skfuzzy.membership as mf
import matplotlib.pyplot as plt

# Variables
# Inputs (height [0m to 5m], dH/dt [-0.225 to 0.225])
x_height = np.arange(0, 5, 0.001)
x_rate = np.arange(-0.25, 0.25, 0.001)

# Output (valve open [0 to 4]: (0) much_less_open, (1) less_open,
(2) no_change, (3) more_open, (4) much_more_open)
y_valve = np.arange(0, 1, 0.001)

print("\n--------------Fuzzy Logic Controller---------------\n")
print("Insert the tank height and the flow rate of liquid. \n")

input_height = float(input("Height [0.0m to 5.0m]: "))
input_rate = float(input("dH/dt (fill rate) [-0.225 to 0.225]: "))

print("\nThe fuzzy logic controller simulation is being performed.\n")

# Membership Functions
# Inputs mf (trapezoidal & triangular mf)
height_high = mf.trapmf(x_height, [2.5, 3, 5, 5])
height_setpoint = mf.trimf(x_height, [2, 2.5, 3])
height_low = mf.trapmf(x_height, [-2, 0, 2, 2.5])

rate_positive = mf.trapmf(x_rate, [0, 0.1, 0.3, 0.3])
rate_zero = mf.trimf(x_rate, [-0.1, 0, 0.1])
rate_negative = mf.trapmf(x_rate, [-1, -0.5, -0.1, 0])
# Output mf (triangular mf)
valve_much_closed = mf.trimf(y_valve, [0, 0.16, 0.32])
valve_closed = mf.trimf(y_valve, [0.16, 0.32, 0.48])
valve_no_change = mf.trimf(y_valve, [0.32, 0.48, 0.64])
valve_open = mf.trimf(y_valve, [0.48, 0.64, 0.80])
valve_much_open = mf.trimf(y_valve, [0.64, 0.80, 0.96])
```

```
# Plots
fig, (ax0, ax1, ax2) = plt.subplots(nrows = 3, figsize =(7, 12))

ax0.plot(x_height, height_high, 'r', linewidth = 2, label = 'High')
ax0.plot(x_height, height_setpoint, 'g', linewidth = 2, label =
'At Set Point')
ax0.plot(x_height, height_low, 'b', linewidth = 2, label = 'Low')
ax0.set_title('Height')
ax0.legend()

ax1.plot(x_rate, rate_positive, 'r', linewidth = 2, label = 'Positive')
ax1.plot(x_rate, rate_zero, 'g', linewidth = 2, label = 'Zero')
ax1.plot(x_rate, rate_negative, 'b', linewidth = 2, label = 'Negative')
ax1.set_title('dh/dt')
ax1.legend()

ax2.plot(y_valve, valve_much_closed, 'r', linewidth = 2, label =
'Much_Less_Open')
ax2.plot(y_valve, valve_closed, 'g', linewidth = 2, label = 'Less_
Open')
ax2.plot(y_valve, valve_no_change, 'b', linewidth = 2, label =
'No_Change')
ax2.plot(y_valve, valve_open, 'y', linewidth = 2, label = 'More_Open')
ax2.plot(y_valve, valve_much_open, 'm', linewidth = 2, label = 'Much_
More_Open')
ax2.set_title('Output Membership Function')
ax2.legend()

plt.tight_layout()
plt.savefig('mf_inputs_output.png', dpi=500)

# Fuzzification
height_fit_high = fuzz.interp_membership(x_height, height_high,
input_height)
height_fit_setpoint = fuzz.interp_membership(x_height, height_
setpoint, input_height)
height_fit_low = fuzz.interp_membership(x_height, height_low,
input_height)

rate_fit_positive = fuzz.interp_membership(x_rate, rate_positive,
input_rate)
rate_fit_zero = fuzz.interp_membership(x_rate, rate_zero, input_rate)
rate_fit_negative = fuzz.interp_membership(x_rate, rate_negative,
input_rate)


# Base Rules
rule1 = np.fmin(np.fmin(height_fit_high, rate_fit_positive), valve_
much_open)
rule2 = np.fmin(np.fmin(height_fit_high, rate_fit_zero), valve_
much_open)
rule3 = np.fmin(np.fmin(height_fit_high, rate_fit_negative), valve_
open)
```

```
rule4 = np.fmin(np.fmin(height_fit_setpoint, rate_fit_positive),
valve_open)
rule5 = np.fmin(np.fmin(height_fit_setpoint, rate_fit_zero), valve_
no_change)
rule6 = np.fmin(np.fmin(height_fit_setpoint, rate_fit_negative),
valve_closed)
rule7 = np.fmin(np.fmin(height_fit_low, rate_fit_positive), valve_
closed)
rule8 = np.fmin(np.fmin(height_fit_low, rate_fit_zero), valve_much_
closed)
rule9 = np.fmin(np.fmin(height_fit_low, rate_fit_negative), valve_
much_closed)

## Union Sets (Mamdani)
v_much_open = np.fmax(rule1,rule2)
v_open = np.fmax(rule3,rule4)
v_no_change = np.fmax(rule5,rule5)
v_closed = np.fmax(rule6, rule7)
v_much_closed = np.fmax(rule8, rule9)


# Plots
valve0 = np.zeros_like(y_valve)

fig, ax0 = plt.subplots(figsize = (7, 4))
ax0.fill_between(y_valve, valve0, v_much_open, facecolor = 'm',
alpha = 0.7)
ax0.plot(y_valve, valve_much_open, 'm', linestyle = '--')
ax0.fill_between(y_valve, valve0, v_open, facecolor = 'y', alpha = 0.7)
ax0.plot(y_valve, valve_open, 'y', linestyle = '--')
ax0.fill_between(y_valve, valve0, v_no_change, facecolor = 'b',
alpha = 0.7)
ax0.plot(y_valve, valve_no_change, 'b', linestyle = '--')
ax0.fill_between(y_valve, valve0, v_closed, facecolor = 'g',
alpha = 0.7)
ax0.plot(y_valve, valve_closed, 'g', linestyle = '--')
ax0.fill_between(y_valve, valve0, v_much_closed, facecolor = 'r',
alpha = 0.7)
ax0.plot(y_valve, valve_much_closed, 'r', linestyle = '--')
ax0.set_title('Valve')

plt.tight_layout()
plt.savefig('output.png', dpi=500)


# Defuzzification
out_valve = np.fmax(np.fmax(np.fmax(np.fmax(v_much_open, v_open),
v_no_change), v_closed), v_much_closed)
#out_valve = np.fmin(np.fmin(np.fmin(np.fmin(v_much_open, v_open),
v_no_change), v_closed), v_much_closed) #doesn't work

defuzzified  = fuzz.defuzz(y_valve, out_valve, 'centroid')
```

```
print("Valve Status :", defuzzified)

# Plot
fig, ax0 = plt.subplots(figsize=(7, 4))

ax0.plot(y_valve, valve_much_closed, 'r', linewidth = 0.5, linestyle =
'--')
ax0.plot(y_valve, valve_closed, 'g', linewidth = 0.5, linestyle =
'--')
ax0.plot(y_valve, valve_no_change, 'b', linewidth = 0.5, linestyle =
'--')
ax0.plot(y_valve, valve_open, 'y', linewidth = 0.5, linestyle = '--')
ax0.plot(y_valve, valve_much_open, 'm', linewidth = 0.5, linestyle =
'--')

ax0.fill_between(y_valve, valve0, out_valve, facecolor = 'Orange',
alpha = 0.7)
ax0.plot([defuzzified , defuzzified], [0, result], 'k', linewidth
= 1.5, alpha = 0.9)
ax0.set_title('Centroid Defuzification')

plt.tight_layout()
plt.savefig('output_defuzzified.png', dpi=500)
```

APPENDIX B

Fuzzy Algorithm for Predicting Power Demand in the Example of Chapter 5 (in Python)

```
#Import the needed libraries
import numpy as np
import skfuzzy as fuzz
from skfuzzy import control as ctrl
import matplotlib.pyplot as plt

# Define the universe of discourse of the input and output variables
time_of_day = ctrl.Antecedent(np.arange(0, 24, 1), 'time of day')
temperature = ctrl.Antecedent(np.arange(40, 111, 1), 'temperature')
electricity_price = ctrl.Antecedent(np.arange(40, 86, 1),
'electricity price')
power_demand = ctrl.Consequent(np.arange(4000, 8001, 1), 'power
demand')

#Membership functions of the input and output variables
time_of_day['sleep'] = fuzz.trapmf(time_of_day.universe, [0, 0, 4, 8])
time_of_day['work'] = fuzz.trapmf(time_of_day.universe, [6, 9, 14,
17])
time_of_day['leisure'] = fuzz.trapmf(time_of_day.universe, [16, 19,
23, 23])
```

```
temperature['cold'] = fuzz.trapmf(temperature.universe, [40, 40,
50, 56])
temperature['cool'] = fuzz.trimf(temperature.universe, [51, 57, 65])
temperature['comfortable'] = fuzz.trimf(temperature.universe,
[64, 70, 77])
temperature['warm'] = fuzz.trimf(temperature.universe, [76, 80, 85])
temperature['hot'] = fuzz.trapmf(temperature.universe, [81, 93,
110, 110])

electricity_price['off-peak'] = fuzz.trimf(electricity_price.
universe, [40, 40, 55])
electricity_price['normal'] = fuzz.trapmf(electricity_price.
universe, [50, 60, 70, 80])
electricity_price['peak'] = fuzz.trapmf(electricity_price.universe,
[75, 80, 85, 85])

power_demand['low'] = fuzz.trimf(power_demand.universe, [4000,
4000, 4500])
power_demand['medium'] = fuzz.trimf(power_demand.universe, [4250,
5400, 6500])
power_demand['high'] = fuzz.trapmf(power_demand.universe, [6000,
6500, 8000, 8000])

# You can see how these look with .view()
time_of_day.view()
temperature.view()
electricity_price.view()
power_demand.view()

#Rules
rule1 = ctrl.Rule(time_of_day['work'] & temperature['cold'] &
electricity_price['normal'], power_demand['high'])
rule2 = ctrl.Rule(time_of_day['sleep'] & temperature['cool'] &
electricity_price['off-peak'], power_demand['low'])
rule3 = ctrl.Rule(time_of_day['leisure'] & temperature['comfortable']
& electricity_price['normal'], power_demand['medium'])

#Create the set of rules and give values to inputs
pd_ctrl = ctrl.ControlSystem([rule1, rule2, rule3])
pd = ctrl.ControlSystemSimulation(pd_ctrl)
pd.input['time of day'] = 13
pd.input['temperature'] = 45
pd.input['electricity price'] = 58

#Defuzzification and final output
pd.compute()
print (pd.output['power demand'])
power_demand.view(sim=pd)
```

APPENDIX C Review Questions

Chapter 1 Review Questions

1. What is the difference between syntax and semantics?
2. How can one describe fuzziness?
3. How are fuzzy sets different from crisp sets?
4. How is the characteristic function translated for infinite valued logic?
5. How are fuzzy logic and logic programming related to each other?
6. What are some examples of logical operations that can be used to formulate propositions?
7. What are some guidelines that act as metrics to create actionable, corrective and interpretable AI systems?
8. Why is the ability to generalize so important?

Chapter 2 Review Questions

1. What is fuzziness in mathematical terms?
2. What is called a "singleton"?
3. How would you describe a membership function?
4. How are *min/max* operators defined?
5. When is a fuzzy set called empty, and when is it called normal?
6. Describe the union and the intersection of two fuzzy sets A and B.
7. What is an α-cut of a fuzzy set?
8. Describe the complement of a fuzzy set.

Chapter 3 Review Questions

1. What is the importance of an *if/then* rule?
2. How can *if/then* rules be formulated?
3. What are the different representations of relations?
4. How are fuzzy relations different from crisp relations?
5. What is the identity fuzzy relation R_I?
6. What is the universe relation R_E?
7. What is the null relation R_0?
8. How is an inverse fuzzy relation expressed?
9. How are implication operators used?
10. What are *GMT* and *GMP*?

Chapter 4 Review Questions

1. Which can be considered as input variables in fuzzy control?
2. Why smoothing of the values of the variables may be used?
3. What is scaling used for?
4. When can quantization of the values of the variables be used?
5. Which are the defuzzification methods?
6. Which defuzzification method does not consider the overall shape of the fuzzy output $\mu_{OUT}(u)$?

Chapter 5 Review Questions

1. Why is forecasting so important?
2. How can fuzzy logic be used for forecasting?
3. How can fuzzy logic be used for data preprocessing?
4. How can one use probability modeling for forecasting?

Chapter 6 Review Questions

1. What is the importance of *approximate reasoning* proposed by Zadeh?
2. How are statistical approaches different from logical approaches?
3. How can fuzzy logic model uncertainty?
4. Why is uncertainty modeling preferred in some cases?
5. How can fuzzy controllers be implemented?

Homework Problems

Chapter 2 Problems

Problem 1

Given fuzzy sets A and B

$$A = 0.3/6 + 0.7/7 + 1.0/8 + 0.6/9 + 0.2/10$$
$$B = 0.2/4 + 0.5/5 + 1.0/6 + 0.6/7 + 0.3/8$$

a) What is the union of the two sets, that is, write an expression for $A \vee B$ (union)?

b) Write an expression for $A \vee B$ (intersection).

c) What is the negation of A?

Problem 2

The set A below describes the category *small integers*,

$$A = 1.0/1 + 1.0/2 + 0.75/3 + 0.5/4 + 0.3/5 + 0.1/7 + 0.1/8$$

a) What is the opposite of *small integers* (under some conditions it can be thought of as the antonym of *small integers*)?

b) What is the category very *small integers*?

c) What is the category *not-so-small integers*?

Sketch Zadeh diagrams for all sets.

Problem 3

The set A has membership function,

$$\mu_A(x) = \frac{1}{1 + \left(\frac{x}{5}\right)^3}, \quad x \geq 0$$

a) If A stands for the category *small numbers*, what is the *concentration* of A and what does it stand for?

b) What is the *dilation* of *small numbers*?

c) What is the category *not small numbers*?

Sketch Zadeh diagrams for all sets.

Problem 4

Discuss and illustrate using the two values of Problem 1, how the "resolution principle" can represent the membership of a fuzzy number. Graph your results for the values $\alpha = 0, 0.25, 0.5, 0.75, 1.0$

Problem 5

If the fuzzy set A in Problem 1 represents the concept of the neighborhood of the number 8, that is, "*around 8*" show what the fuzzy set of the concept "*very much around 8*" and "*more or less around 8.*" Repeat for the set B in problem 1 which can be thought of as "*around 6.*"

Problem 6

The fuzzy sets, which are normal and convex, can represent fuzzy numbers A and B given by

$$A = 0.33/6 + 0.67/7 + 1.00/8 + 0.67/9 + 0.33/10$$

$$B = 0.50/1 + 1.00/3 + 0.50/5$$

a) How do we know that they are fuzzy numbers? In other words, show that they are normal and convex.

b) Assume that the membership functions for these fuzzy numbers are triangular and continuous and that the lines of the triangles go through the points indicated above in A and B above. Use the extension principle to subtract B from A to give a fuzzy number C. Sketch A, B, and C memberships in Zadeh diagrams.

Problem 7

For the fuzzy set A in Problem 4, determine the following α-cuts:

a) $A_{0.8}$

b) $A_{0.6}$

c) $A_{0.3}$

Problem 8

Use the extension principle to fuzzify the straight line given by the equation $5x + 2y + 10 = 0$. Include any necessary assumptions.

Problem 9

Consider the following two fuzzy sets:

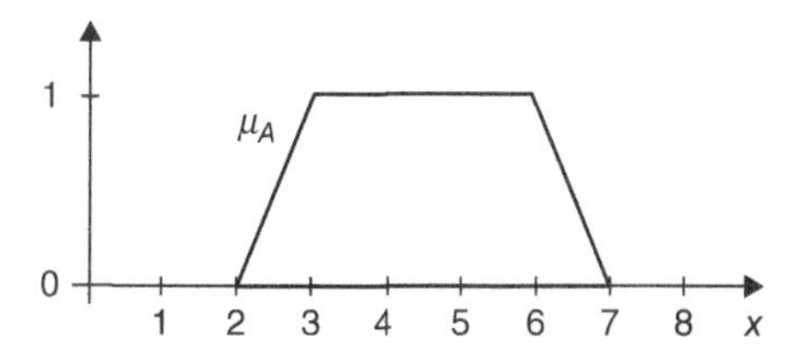

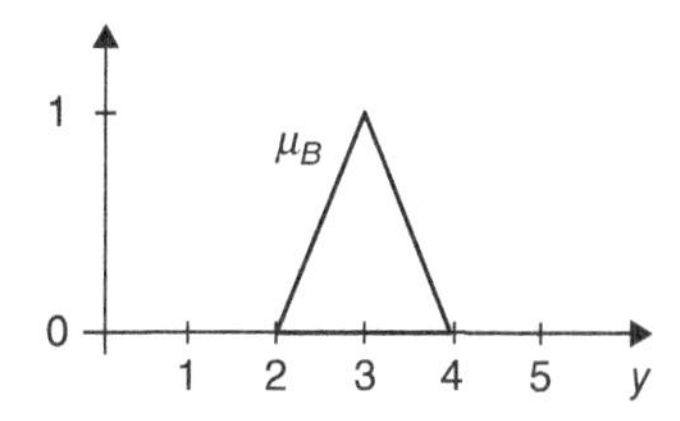

Use the extension principle to apply the following two functions (mappings) to these fuzzy sets and find the fuzziness of z induced by the mappings below:

a) $z = \dfrac{1}{x}$

b) $z = x - 2y$

Chapter 3 Problems

Problem 1

Using *Mamdani implication operator*, if we have

$$A \rightarrow B \equiv A \wedge B$$

Show that the following is also true:

$$\begin{aligned} A_1 \rightarrow A_2 \rightarrow B &= A_1 \wedge (A_2 \wedge B) \\ &= (A_1 \wedge A_2) \wedge B \\ &= (A_1 \wedge A_2) \rightarrow B \end{aligned}$$

Problem 2

If *Boolean implication* is used, and have the rule,

$$A \rightarrow B \equiv \bar{A} \vee B,$$

Show that the following is true:

$$\begin{aligned} A_1 \to A_2 \to B &= A_1 \to (\bar{A}_2 \vee B) \\ &= \bar{A}_1 \vee (\bar{A}_2 \vee B) \\ &= (\bar{A}_1 \vee \bar{A}_2) \vee B \\ &= \overline{(A_1 \wedge A_2)} \vee B \\ &= (A_1 \wedge A_2) \to B \end{aligned}$$

Problem 3

A linguistic description is comprised of a single rule

$$\textit{if } x \textit{ is } A \textit{ then } y \textit{ is } B$$

where A and B are the following fuzzy values

$$A = 0.33/6 + 0.67/7 + 1.00/8 + 0.67/9 + 0.33/10$$
$$B = 0.50/1 + 1.00/3 + 0.50/5$$

Use *Larsen product implication operator* for the rule. If a fuzzy number *x is A′* is a premise, use *generalized modus ponens* to infer a fuzzy number *y is B′* as the consequent. A' is defined by

$$A' = 0.5/5 + 1.00/6 + 0.67/7 + 0.33/8$$

Use the *max-min composition*.

Problem 4

Repeat Problem 3 using *Mamdani implication operator* and *max-min composition*. Plot B' and discuss the differences with B' obtained in Problem 3. What accounts for the difference?

Problem 5

Two fuzzy *if/then* rules in a fuzzy algorithm can be merged simplifying in this way the algorithm. The merging can take place in two relations if they differ in no more than one proposition in the *LHS*. For instance,

$$\textit{if } A_1 \textit{ then } (\textit{if } B \textit{ then } C)$$
$$\textit{if } A_2 \textit{ then } (\textit{if } B \textit{ then } C)$$

Show that it is true that

$$\textit{if } (A_1 \textit{ OR } A_2) \textit{ then } (\textit{if } B \textit{ then } C)$$

Problem 6

Two fuzzy *if/then* rules in a fuzzy algorithm can be merged to simplify the algorithm. Given the two rules

$$\text{if } A \text{ AND } B_1 \text{ then } C$$
$$\text{if } A \text{ AND } B_2 \text{ then } C$$

Show that they can be merged into a single rule of the form

$$\text{if } A \text{ AND } (B_1 \text{ OR } B_2) \text{ then } C$$

Problem 7

Combining n fuzzy fuzzy *if/then* rules in a cluster for which

$$RC^N = \bigvee_{j=1}^{n} R^j$$

What kind of implication relation and connective can give for the rule cluster membership function

$$\mu_{R^N}(x,y) = \bigvee_{j=1}^{n} (\mu_{A^j}(x) \wedge \mu_{B^j}(y))$$

Show all assumptions and operations.

Problem 8

The combination of n fuzzy *if/then* rules in a cluster,

$$RC^N = \bigwedge_{j=1}^{n} R^j$$

gives as membership

$$\mu_{R^N}(x,y) = \bigwedge_{j=1}^{n} [1 \wedge (1 - \mu_{A^j}(x) + \mu_{B^j}(y)]$$

Determine the *implication operator* and *composition* that would produce this membership function.

Problem 9

Consider the fuzzy *if/then* rule

$$\text{if } x \text{ is } A \text{ then } y \text{ is } B$$

Where the two fuzzy are given by the sets

$$\{\mu_A(x)\} = \{0, 0.2, 0.7, 1.0, 0.4, 0\}$$
$$\{\mu_B(y)\} = \{0.3, 0.8, 1.0, 0.5, 0\}$$

Where the universes of discourse X, Y are

$$X = Y = \{1, 2, 3, 4, 5, 6\}$$

If *Zadeh implication* is used, find the fuzzy relation of the rule.

Chapter 4

Problem 1

The fuzzy sets below are outputs of two rules of a controller at some time step

$$b^1 = 0.33/6 + 0.67/7 + 0.5/8 + 0.67/9 + 0.33/10$$
$$b^2 = 0.50/1 + 0.5/3 + 0.50/5$$

Assume triangular shapes (for the original *RHS* of the rules involved) and Mamdani implication.

1. What is the compound output (use Zadeh diagrams)?
2. What is the defuzzified output if *COA* defuzzification is used?
3. What is the defuzzified output if *MOM* defuzzification is used?

Problem 2

The fuzzy sets A and B are in an *if/then* rule of a controller

$$\textit{if } x \textit{ is } A \textit{ then } y \textit{ is } B$$

where,

$$A = 0.3/6 + 0.7/7 + 1.0/8 + 0.6/9 + 0.2/10$$
$$B = 0.2/4 + 0.5/5 + 1.0/6 + 0.6/7 + 0.3/8$$

Assume triangular shapes (for the original *LHS* and *RHS* of A and B, respectively) and Boolean implication.

Show and sketch what the degree of fulfillment (*DOF*) of the rule is for the singletons below given as inputs to the rule,

a) $A' = 1.0/6$

b) $A' = 1.0/7$

c) $A' = 1.0/8$

Problem 3

Repeat Problem 2 with the Mamdani min implication operator.

Problem 4

Repeat Problem 2 with the Larsen product implication operator.

Problem 5

Repeat Problem 2 with the Arithmetic implication operator.

Problem 6

Repeat Problem 2 with the Arithmetic implication operator.

Problem 7

Show analytically how Eq. (4.4) is obtained. State clearly all assumptions made to achieve this result.

Problem 8

Show analytically how the property of distributivity leads to Eq. (4.5) is obtained. State all assumptions made.

Problem 9

Show how for a composite situation $s_{i1} \times s_{i2} \times \cdots \times s_{im}$ and a specific rule j, the degree of fulfillment DOF_i^j will be the same for all the subsets of rules in a rule cluster *RC* and hence it only needs to be computed once for all rules, as shown in Eq. (4. 10).

Problem 10

How would the results of Example 6.1 be different if a *MOM* defuzzification method is used. Compare with *COA* and discuss your findings.

Problem 11

In Example 6.1, use *SOS* defuzzification and compare it with *COA* and your findings by using *MOM* defuzzification in Problem 10.

Chapter 5

Problem 1

List and describe the main five steps in the process of fuzzy inferencing.

Problem 2

Consider the following fuzzy if-then rule: (i) if x is A_1 then y is B_1, (ii) if x is A_2 then y is B_2, where $A_i \in X$ and $B_i \in Y (i = 1,2)$ are fuzzy sets as follows:

$$A_1 = 1/x_1 + 0.75/x_2 + 0.15/x_3$$
$$A_2 = 0.75/x_1 + 1/x_2 + 0.1/x_3$$
$$B_1 = 1/y_1 + 0.1/y_2$$
$$B_2 = 0.1/y_1 + 0.8/y_2$$

Given the fact that x is A' where $A' = 0.8/x_1 + 0.75/x_2 + 0.15/x_3$, use the composition rule of inference to calculate the conclusion B'.

Problem 3

Solve the short-term electricity demand forecasting problem showed in Chap. 5.2, using the following defuzzification methods:

1) Center of area (COA), i.e., centroid, and
2) Mean of maxima (*MOM*).

Use the same variables, fuzzy sets, and membership functions.

Problem 4

Show how the transition diagram of Fig. 5.7 is obtained in the forecasting with the probability model.

Problem 5

Using Python (or a language of your choice) and weather data from your area to construct a transition diagram predicting the next day temperature. Obtain a similar transition diagram to the one used in Fig. 5.7.

Problem 6

Use weather data from your area to develop a simple fuzzy algorithm predicting the next day temperature and compare your results with the probability model of Problem 5. You may use any language of your choice, including Python or the MATLAB Fuzzy Toolbox.

Chapter 6

Problem 1

A fuzzy control system for PUR-1 uses inputs of error e and change in error Δe to control an output variable u (Control Rod Height). Their fuzzy membership functions and fuzzy algorithm are given below. Determine the output u (%) for $e = +10\%$ and $\Delta e = -2\%/min$ using *Mamdani implication* and *max–min composition*. Use the *Center of Area* method to defuzzify the answer. Sketch the various membership functions involved and to show how you obtained your solution.

The fuzzy algorithm is as follows

R_1 *if e is N and* Δe *is N, then u is L, ELSE*
R_2 *if e is N and* Δe *is P, then u is M, ELSE*
R_3 *if e is P and* Δe *is N, then u is M, ELSE*
R_4 *if e is P and* Δe *is P, then u is S*

The fuzzy values used have Zadeh diagrams as shown below,

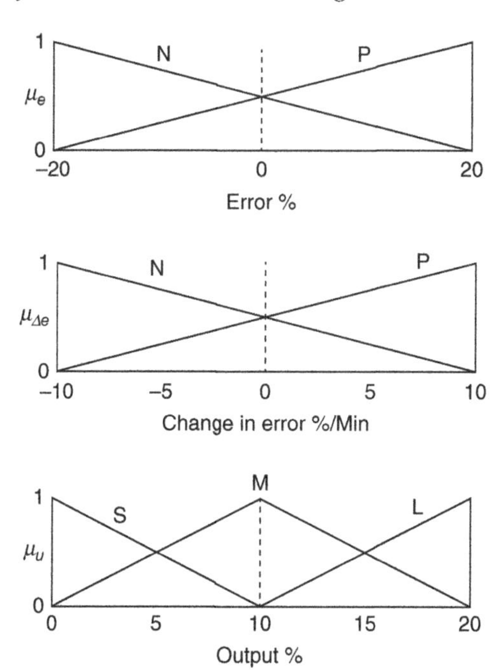

Problem 2

Create a compressor fan speed fuzzy controller system using the five-step method:

1. Fuzzify Inputs
2. Apply Fuzzy Operator, *hint*: Since there are no ANDs or ORs in the rules, no fuzzy operator is needed for step 2.
3. Apply Implication Method, *hint*: Use the MIN implication method.
4. Aggregate All Outputs using the MAX method.
5. Defuzzify using the CENTROID method.

There are two fuzzy rules:

*if Temperature is **HOT** then turn fan speed to **HIGH***
*if Temperature is **MEDIUM** then turn fan speed to **MEDIUM***
*if Temperature is cold then turn fan speed to **LOW***

i. Construct a system to find the crisp output for temperatures = 1 to 100. Plot the input output relationship and explain any peculiarities.

Define universe of discourse and membership functions:
Input universe of discourse is from 0 to 100.
Use gaussian temperature membership functions with the following centers and spreads that let the MFs overlap at around the 0.5 level.

hot parms => center = 0
medium parms => center = 50
cold parms => center = 100

Output universe of discourse is from 0 to 10.
Use triangular fan speed membership functions (mf_tri([a c b],'n').

high parms => [5 10 10]
medium parms => [2 5 8]
low parms => [0 0 5]

ii. Write it as an m-file script, comment the code fully, and turn in the m-file and a diary of its use.

iii. Write up a short document showing and explaining its operation, use necessary graphs.

Index

N

U

V

W